THE Kindly DR. Guillotin

HAROLD J. MOROWITZ

THE Kindly
DR. Guillotin

and Other Essays

on Science and Life

COUNTERPOINT
WASHINGTON, D.C.

Many of these essays were previously published in the journal
Hospital Practice, and are reproduced with permission of The
McGraw-Hill Companies.

Library of Congress Cataloging-in-Publication Data
Morowitz, Harold J.
    The kindly Dr. Guillotin : and other essays on science and life /
    Harold J. Morowitz.
    1. Science. I. Title
Q113.M74 1997
081—dc21   97–28905

ISBN 1-887178-49-X (alk. paper)

Book design and electronic production by David Bullen Design

Printed in the United States of America on acid-free paper that
meets the American National Standards Institute Z39-48 Standard

COUNTERPOINT
P. O. Box 65793
Washington, D. C. 20035-5793

Distributed by Publishers Group West

10 9 8 7 6 5 4 3 2 1

FIRST PRINTING

To the wife of my youth,
who still abides with me

CONTENTS

Preface

Writing essays can be a serious or lighthearted endeavor, but gathering these writings into a collection almost always becomes a serious matter. For one must then sort the individual pieces into groupings—or into *skandas* that, according to the Buddhist worldview, relate our vanities to the idea of self. In this process of sorting, local thoughts can be set within more global categories. Thus I have grouped these essays into "People and Places," "Language," "Science," "The Ecosystem," "Criticism," and "Commentary."

"People and Places" resonates with the skanda of *feeling*: my response to individuals and locations that have touched or puzzled me. There is no predicting these interactions. They occur along the road of life, and one simply reports on them and muses on the experience.

The second section, "Language," corresponds to the skanda of *form*. The pieces deal with intriguing words and phrases (language in the small) and with literature (language in the large).

These matters are at the core of what a writer must comprehend, intuit, and wrestle with. To many editors language is mostly a matter of form; to many authors it is largely a matter of voice. This may result in a certain amount of creative tension. I must confess, however, that I find crafting prose more fun than labor.

The third grouping, "Science," is the bone and sinew of my being. It is the framework on which I try to organize thought and understanding. That I don't always succeed at this endeavor speaks to the role of all the other aspects of being. It is not possible to reduce all of experience to the logic of science. I identify science with the skanda of *consciousness,* for the conscious mind is the agent for building theoretical constructs, and these constructs are the stuff of scientific understanding.

The "Ecosystem" section brings together a subset of the science essays, the part that speaks of the plants and animals and microbes that share our world around us. Its skanda is *perception,* for while all science starts with the observed, the ecosystem is the constant background of all perceptual activity as we go through day-to-day experience.

The fifth category, "Criticism," has relevance to the skanda *impulse.* For criticism is a curmudgeonly thing that responds to the slings and arrows of fortune by denouncing them in one form or another. If the impulse of this section lacks charity, it perhaps seeks for justice: some things just deserve to be criticized.

The final grouping, "Commentary," corresponds to no skanda. It is an omnibus that contains bits of all the other categories.

Yet, following Melville, we note that all great systems of classification must be incomplete. And Melville is, after all, a good author to follow.

H. J. M.

I

People and Places

The Kindly Dr. Guillotin

THE PASSING of the revered Supreme Court Justice Thurgood Marshall, who was firmly convinced that capital punishment is unconstitutional, led me to some old notes and some thoughts about executions and executioners and the man whose name has come to be forever identified with decapitation.

History has not been kind to Joseph Ignace Guillotin. His name is now synonymous with a machine for decapitation that he did not invent, operate, or fall victim to. We have a terrifying image of Doctor Guillotin, who was physician, politician, and humanitarian. Born in Santes in 1739, Guillotin studied medicine and became one of eighteenth-century France's distinguished physicians.

While he was so engaged, he interacted with some of the major scientific and intellectual figures of that period. In 1778, Franz Mesmer came to Paris to demonstrate animal magnetism. Mesmerism, or hypnosis as we might now call it, made such an impression and caused such a stir in Paris that Louis XVI appointed a royal commission to study the phenomenon. Among the members were Antoine Lavoisier, the father of modern chemistry; Benjamin Franklin, widely known in Europe for his studies of electricity; Jean-Sylvain Bailly, noted astronomer; and the physician Guillotin. The commission concluded that there was nothing to Mesmerism—it was all in the mind. Such was the view of the eighteenth-century rationalists.

*A*fter the storming of the Bastille, Guillotin was elected to the National Assembly, and on December 1, 1789, he offered a formal resolution on capital punishment. He proposed that all those to be executed, both commoners and nobility, should be killed by mechanical decapitation. This legislation was merciful from several points of view. Before the Revolution, nobility had the privilege of decapitation, whereas ordinary folk were hanged. In addition, the decapitations were carried out by an executioner using a large ax. There were clumsy executioners, so the use of a machine made the procedure more uniform and thus fairer. Indeed, in the days of manual executions, it was the practice in some places to tip the executioner to assure a good, clean, quick job. No doubt about it: Dr. Guillotin's suggestions were egalitarian and humanitarian in intent.

There was a history of using execution machines to decapitate the convicted. From the thirteenth century onward, a device known as the *Mannaia* was used on nobles in southern Italy. In the Middle Ages in Germany, there was a series of decapitators known as *Diele, Hobel,* and *Dolalora*. There is also such a machine in the Antiquarian Museum of Edinburgh that was used between 1581 and 1685. A similar model was in use in Toulouse in 1632. Thus, Dr. Guillotin is by no means the inventor of mechanical beheading. His proposal became law on October 6, 1791. Guillotin also did not build the decapitator. It appears to have been fabricated in Germany and experimentally tested on human corpses and live sheep at Hôpital Bicêtre. The first published reference to the device as "la guillotine" appears in *Journal des Revolutions de Paris et de Brabant,* April 1792, published by Camille Desmoulins.

During the Terror, Joseph Guillotin was arrested but was spared decapitation and imprisoned instead. He was subsequently released in the Thermidorian reaction and resumed an active public life. He played an influential role in reestablishing the Academy of Medicine and died a peaceful, natural death in 1814.

The machine that bears Dr. Guillotin's name was used for capital punishment until 1977, when the death penalty was abolished in the Republic of France. Nevertheless, the good doctor's name lives on in the noun *guillotine* and verb *to guillotine,* both bloody reminders of a man whose motives were fairness and compassion for the condemned.

t is interesting how people's names enter the language, as in Newton's laws and Darwin's theory of evolution. Some are remembered by units, such as the ampere, the volt, the ohm. A few enter as adjectives, as in kekulé structures and diesel engines. Other notable examples are Maxwell's equations and Gibbs free energy. Boltzmann's constant must represent a high form of recognition. I personally recall reading a journal on photobiology and coming across the "Morowitz correction factor" for ultraviolet dosimetry. I decided that being remembered for a correction factor must be the lowest rung on the ladder of nomenclatural honor, but you take what you can get!

This brings us back to Dr. Guillotin and why the mention of this man's name evokes terror. In the United States, we practice capital punishment by hanging, lethal injection, poisonous gas, electrocution, and shooting. Are these any kinder ways of execution than the machine of Guillotin? They are clearly less bloody, but that seems like an excessively aesthetic criterion when talking about life and death. Why is decapitation, which was favored in the 1700s, now looked on with such abhorrence? I suspect we will never know. The act of a civilized state's taking a human life is so inherently quirky that the discussion of methods also takes on elements of the theater of the absurd.

This returns us to thoughts of the departed Justice Thurgood Marshall. He so opposed capital punishment that he universally held it unconstitutional on grounds of the ban against cruel and unusual punishment. To my knowledge, none of the sitting Supreme Court justices shares this view. Whether or not one agrees with Marshall's ideas, they deserve to be remembered as a message to consider what is at stake in the state's taking a life by whatever method. I suggest that we establish a verb, "to thurgood," that is, to spare a condemned criminal from execution and substitute life imprisonment. Farewell, Justice Marshall.

The Professor

THE PROBLEM with a writer's maturing is the danger of falling into the literary genre of the obituary as one's mentors and colleagues, to use Tennyson's term, "cross the bar." I would be tempted to forgo the greatness of the past and push forward to the greatness of the future, but for two guiding principles: first, the importance of establishing intellectual continuity across the years, and, second, my need to express my profound gratitude for the extreme good fortune that has brought me such a full share of inspired mentors and colleagues.

Thus, the July 23, 1992, newspapers brought news of the passing of F. S. C. Northrop at the age of ninety-eight. The professor was truly a polymath, having written significant works in law, the philosophy of science, and international relations. I suspect he will be remembered best for *The Meeting of East and West,* a work that in 1946 first introduced the American reading public to Oriental philosophy and to the coming impact of Eastern thought on the Western world.

In *Zen and the Art of Motorcycle Maintenance* (1974), Robert M. Pirsig tells of reading *The Meeting of East and West* on a troopship returning from the Korean conflict. Northrop's book was a major component in Pirsig's decision to return to the university to study philosophy. I have the feeling that many individuals were affected to an equal degree by this book.

Professor Northrop had a habit of being prescient. In 1952, he published *The Taming of the Nations: A Study of the Cultural Bases of International Policy*. Chapter Eight, "The Resurgence of Islam," provided a clue to forthcoming events in the Muslim world over the next forty years.

I first met Northrop in 1946 when I was a Yale undergraduate. I enrolled in his logic course and subsequently took his course on the philosophy of science. I must confess that I did not then grasp the full impact of my instructor's words, nor did I appreciate the importance of the scholarship in which he was engaged. I did, however, understand that I was in the presence of a very deep thinker.

I can recall sitting in a classroom in Harkness Hall listening to the professor discuss a favorite theory of his about how new conceptualizations of nature were introduced in science, worked their way into philosophy, and finally emerged in political and social thought. He would lecture on the physics of Newton, its influence on the epistemology of John Locke, and its ultimate effect on the American and French revolutions.

I next encountered the philosopher in 1979 in the office of Henry Margenau, professor emeritus and another mentor from those carefree undergraduate days when I did not fully realize the quality of the education to which I was being treated. We chatted about epistemology, and Northrop told us stories about his interactions with Albert Einstein, Alfred North Whitehead, and Lawrence J. Henderson. Then Margenau added anecdotes about his own meetings with Einstein and Erwin Schrödinger. I was uncharacteristically quiet but fully cognizant of the lesson in historical continuity.

Northrop was among the first of the philosophers to appreciate the impact of twentieth-century science on philosophy. In the 1920s and 1930s, he had made his reputation exploring consequences of the new physics. His view always went beyond the science to ask how these new discoveries would affect art, literature, law, and a wide range of human endeavors.

When I encountered Northrop again in 1984, he was engaged in a four-volume series to be entitled *Philosophiae Naturalis Principia Mathematica et Aesthetica et Jurisprudencia et Summa Theologica*. I thought that the efforts of a ninety-year-old man to begin to assemble such an epic work represented optimism at its very highest. He had been traveling around the world, unable to find a publisher to share his optimism and undertake publication of the series. Now he was back in New Haven to confer with a small publishing company that had agreed to publish *The Prolegomena to a 1985 Philosophiae Naturalis Principia Mathematica*. Through a complicated series of events, I finally became the informal editor of this short one-volume introduction to the major work. I took the professor to visit his old philosophy department colleague, the noted logician Frederic B. Fitch. The two emeriti sipped tea and talked symbolic logic. It was a touching experience.

When I got down to editing *The Prolegomena,* I encountered a number of problems: 1) Dr. Northrop had suffered an illness and was in a nursing home where he could not be disturbed; 2) I did not fully comprehend the text; 3) many of the references were from his memory, which was on the whole remarkable but still slightly imperfect; 4) sections of the book dealt with translational difficulties of Euclid's Greek, Newton's Latin, and Einstein's German; and 5) Professor Northrop would occasionally coin a new word, and on at least one occasion the new word was in Septuagint Greek. However, I plowed ahead, with great editorial guidance from Northrop's scholars Horace Oliver and Norman Riise, and the book actually appeared in 1985.

Owing to Northrop's declining health, *The Prolegomena* was his last work, and we can only speculate about the four-volume series. His plan, as I understand it, was based on his assertion that the physics of Einstein's relativity and Werner Heisenberg's quantum mechanics contains certain epistemological assumptions, which, if expanded and fully understood, would be of enormous value in the humanities and social sciences and also in the search for theological understanding. To him the ideas were deeply rooted in classical Greek and biblical writings, and his mind constantly moved back and forth across the centuries. The seventy-three pages of *The Prolegomena* are the most densely packed prose I have ever encountered. Perhaps its understanding awaits another scholar of Northrop's depth and stature. The book ends with a quotation from Einstein: "I want to know how God created the universe. I am not interested in this or that spectral line or other phenomenon. I want to know God's mind. The rest are details."

This *In Memoriam* looks to the future and to the continuation of the kind of scholarship characteristic of F. S. C. Northrop—the search for universal principles of human understanding.

Rapa Nui Diary

I<small>T'S A LONG WAY</small> from Koobi Fora in Northern Kenya to Rapa Nui—Easter Island—in the Pacific Ocean. The former is believed by many to be the birthplace of *Homo habilis,* a hominid thought to be a direct ancestor of *Homo erectus,* the progenitor of *Homo sapiens*. The latter site, now beneath my feet, is the place where a small group of islanders of unknown or uncertain origin carved and mounted huge stone statues quarried from the lava rock of the volcano at Rano Raraku. Another group of islanders, for reasons still unknown, pushed over most of the statues, or moai, which now lie face down around the island. I am trying in my own mind somehow to establish the nexus between Koobi Fora and Rapa Nui.

Thor Heyerdahl and his coworkers appear to have solved, in part, the engineering aspects of how these giant figures were constructed and moved. It is impressive technology, but as one stands before a giant figure, the question of *how* is overridden by the question of *why*. Why would the relatively small population of this most isolated island in the world have devoted so much of its time, labor, and productivity to the creation of these totems? I suspect that the answer might provide insight into the emergence of the human mind at Koobi Fora or at some time later in the social development of humanity.

The thought now coming into focus has been building in my mind for many years. It began some time back at Machu Picchu as I stood at dawn staring at structures held together by precisely fitted big, indeed often huge, rocks. It grew while I crawled through the pyramids of Egypt and stood in awe in the great cathedrals of Europe. I get the same feeling from giant nuclear accelerators, from religious art of the Italian Renaissance, and from the study of quantum mechanics.

But at Easter Island the experience of looking at the statues is so intense, so isolated from the rest of the world, and so awesome that one cannot turn away without at least a feeble stab at an explanation. This island could never have supported a population of more than four thousand or five thousand humans (at present there are about two thousand). Only a fraction of the islanders were statue builders. The rest were iconoclasts, in the traditional sense of the word, who eventually tore down the statues and were reported to have eaten some of the statue builders. This iconoclasm reminds me of the many torn-down Lenins in the former Soviet Union and the rest of Eastern Europe.

What is becoming so focused for me on this lonely, isolated volcanic island is the extent to which our humanness has a theoretical component that leads us to actions that make no sense from the point of view of our survival as animals in a Darwinian world of fitness. From the perspective of classic economics or anthropology, human needs are food, shelter, clothing, and procreation. In a Darwinian sense, these will satisfy our success as animals, members of the species *Homo sapiens*.

There must, however, be another component—the drive to theorize. For without an abstract view of some unseen or unsensed forces, there would be no point in constructing idols. Time spent in building these moai is time that is not devoted to improving the material conditions of a society. The reasons for this work then are presumably aesthetic or religious. Again, there is some component that wishes to understand or respond to the unseen. This involves postulating the nature of the unobserved universe, which is, of course, the role of theory.

The pyramids of Egypt were the result of theories held by the ruling class about life after death. The ruling class was able to use the slave class and the resources of that society to accomplish its selfish religious—theoretical—goals. Similarly, the great stone works of Machu Picchu were parts of a religious complex that the Incas had built for them by the enslaved inhabitants of the Andes.

The theoretical component also characterizes Western societies. The great cathedrals of Europe were a response to an unseen God and to a view about that God's role in the world. The expenditure of resources was again significant, but since the theory postulates eternal rewards, the price seemed small.

Theory need not refer solely to religion. For example, consider the supercollider that was under construction in Waxahachie, Texas. This was a project that was going to cost $11 billion or more to answer questions in theoretical physics. When Congress was considering terminating the project, one of the last arguments offered by scientists was the potential of the collider to answer questions about the origin of the universe. The thrust was not only theoretical but philosophical in the broadest sense. I suppose that the enormous filled-in hole in the ground in Texas will stand (or not stand) with the fallen statues of Easter Island as monuments to theoretical structures, those who erect them, and those whose perspective argued for destroying them.

If one assumes that the drive to theory is part of the fundamental description of humankind, then an important question remains as to whether it emerged with the evolution of *Homo sapiens* or one of the previous hominids or whether it is a social construct that has developed with the evolution of *Homo sapiens* societies. On reflection, I am not sure that this question can be well defined. There are cave painting and burial sites with artifacts to suggest theories about death going back fifty thousand years. Theories seem to characterize all cultures that have been subject to anthropologic analysis. Perhaps language always involves a level of theory necessary to reduce the world to symbolic representations.

Well, enough of standing around here and staring in wonderment at these statues. The time has come to bump over the unpaved roads that will take me to many more statues around the island. Indeed, I have heard that there are two thousand moai on Rapa Nui. From what I have seen so far, that seems like a reasonable estimate. This large number only reinforces my argument as to how important these totems were in the lives of some of the inhabitants of this isolated outpost of humankind.

A New Vista
in Natchitoches

T HE AIRPLANE sets down softly on the runway in Shreveport, and a few minutes later I am met by my host, a faculty member of the Louisiana Scholars College at Northwestern State University of Louisiana. The drive, about an hour from Shreveport to Natchitoches, is through soft, green, flat fields. The broad highway divider is full of red spring flowers. Along the roadside are David Duke posters left over from the last election.

Natchitoches is a timeless small town dotted with churches, stores, gas stations, and ageless wooden houses built over a period of almost three centuries. It is the oldest permanent European settlement in Louisiana, dating back to Fort St. Jean Baptiste and the struggle between the French and the Spanish for control of the territory. I have come here to discuss the core science program at Scholars College.

I leave my luggage at the Fleur de Lis Bed and Breakfast. This establishment, a large old frame house, frequently provides lodging for the university's guests. The quality and quantity of the host's breakfasts have a reputation far and wide. I am getting ready for my first meeting, dinner with some students and faculty. And here I have a confession. Although I have read the impressive literature about Scholars College, I am not prepared to find that one of the great undergraduate educational experiments in the United States is taking place in Natchitoches, Louisiana.

The program can be briefly described. About half of the classes are a core common to all students. The "texts and traditions" sequence explores writings from the *Bhagavad Gita* to William Faulkner and Toni Morrison. While focusing on "past classics" (how do we know a contemporary work is a classic?), the curriculum is sensitive to race and gender in imaginative ways— such as reading the autobiography of Frederick Douglass or "On the Subjugation of Women" by John Stuart Mill. Another sequence deals with the sciences in an integrated manner. Tools such as foreign language, critical reading and writing, and computer use are included. Each student then takes an individualized cluster of courses constituting a specialization that culminates in a senior thesis.

My first interaction with a group of faculty and students impresses me with the élan and enthusiasm for the program that they share. Scholars College, only five years old, was founded by University President Robert Alost. It is composed of a group of professors recruited from Northwestern State University and around the country. The faculty brings a level of commitment to their teaching that I have rarely seen in academia in the United States.

The next day I spend conversing, mostly with seniors in small groups of up to five. They are articulate, elegantly civil, and obviously well educated. The subjects of their theses range from organic synthesis to film criticism. Listening to the presentation of two of these projects, one is impressed with the high quality. I am trying to figure out what makes this institution work so well, and that is not so easy to do.

The first thing that engages me is the amount of time that the faculty devotes to the program. For some of them, I would describe it as a calling rather than a job. Second, most faculty members live in Natchitoches, a few minutes away from the college. This allows them to function as resident tutors, somewhat in the fashion of the classical tradition of Oxford and Cambridge described by C. P. Snow. Third, the college has been able to attract faculty members who share the vision. This is a group of people for whom education is very important. In most major American universities, the publish-or-perish ethic leads faculty to a primary loyalty to their national professional peer group rather than to their teaching institution. Many here seem to march to the beat of a different drummer, although they, too, maintain scholarly activity.

In discussing programs such as that of Scholars College with colleagues, I am sometimes reminded that the programs are elitist. I agree but don't understand why the issue is raised. All achievement is elitist. Pole-vaulting is elitist, the university orchestra is elitist, basketball is elitist—why is the issue raised only when the elitism is scholarly and perhaps most tied to the competitive future of the country? Why object to elitist academic endeavors?

I would argue that state universities have two general missions. The first is broad general education and career training for the constituency that wants that type of program. It is also in the interest of the nation, the state, and the individual to have an intense program for gifted scholars who wish to maximize the depth and breadth of their education. Each university should, by this criterion, have a Scholars College, and Northwestern State University of Louisiana is to be commended for its leadership role.

My two and a half days in Natchitoches are full of hard work, discussions of science curriculum, and intense interaction with students and faculty. I am trying to put together the take-home lesson before I leave this land of steel magnolias, but it doesn't quite gel.

On Sunday morning, I'm on the early flight from Shreveport to Atlanta. My companion in the next seat is a New England academic who has been discussing the humanities program at Scholars College. Our conversation provides an opportunity to sort out features of the program and to think about some of the problems as well as the achievements. Even in the light of day, the achievements look impressive.

This decade has been a time to view with alarm the educational system in the United States. I begin to wonder how many other Scholars Colleges there are around the nation. Have we been looking in the wrong places and asking the wrong questions? I would suggest that Harvard, Yale, Stanford, and their ilk take a good look at what's going on in Natchitoches. Professors who enjoy teaching and students who enjoy learning may just be the wave of the future in American education.

Piltdown Man, Harvard Man

A STRANGE CREATURE issued forth from a gravel pit in the Sussex district of England in 1912. In a dig that was being supervised by Charles Dawson, an amateur student of paleontology, the fossils unearthed included parts of a skull and jawbone that were interpreted by many as remains of an early hominid, a missing link, as it were, between ape and man. This postulated human ancestor was given the name Piltdown Man, after the town in which it was discovered. Along with these fossils were found other assorted remains of animals and presumed hominid artifacts.

Piltdown Man, with a contemporary humanlike skull and an apelike jaw, generated great dispute about the course of human evolution. The reconstructed head was at variance with other

finds on prehuman hominids. This virtual hominid lived until
1953, when Kenneth P. Oakley showed by chemical evidence
that the Piltdown finds were part of an elaborate hoax of treated
bones planted to confuse paleontologists.

Following Oakley's proof of fraud, there has been a long series
of speculations as to the identity of the hoaxer. Indeed, it has
become something of a parlor game to try to identify the perpetrator. The more famous and unlikely the presumed hoaxer, the
more points would be earned by the identifier. I find this a grim
game, for the hoax was a vile act, the ultimate felony of a scholar,
creating deliberately false data. Accusing someone of the hoax
would ruin his reputation and thus would require proof beyond
reasonable doubt.

After more than thirty years of such speculations, the May 23,
1996, issue of *Nature* brought the news that the contents of a
trunk found in London's Natural History Museum and carefully
studied by Brian Gardiner, Professor of Palaeontology at Kings
College, London, have provided an almost certain identification
of the perpetrator. The trunk, which belonged to Museum staff
member Martin A. C. Hinton, contained materials treated by the
same method of artificial aging used in the Piltdown materials.
The trunk was a chemistry set for hoaxing. *Nature* notes, "Gardiner feels that the evidence for Hinton having been the sole
hoaxer is now conclusive."

Ah, I thought, the smoking gun. I will not have to worry
about this issue any more. My euphoria lasted only from May to
September, when *The New York Times* brought a review of *Unrav-
eling Piltdown: The Science Fraud of the Century and Its Solution* by

John E. Walsh. Walsh does not agree with Gardiner and is absolutely certain that the hoaxer was Charles Dawson, acting alone.

Now, I must confess that I do not lose sleep worrying about who performed the dastardly Piltdown hoax. But I *have* taken to wondering about people who devote enormous effort to sifting through a large body of circumstantial evidence in an attempt to nail the hoaxer.

Indeed, I am especially reminded of a troubling article in the August 1980 issue of *Natural History* by Harvard professor Stephen Jay Gould, entitled "The Piltdown Conspiracy." In this long, detailed essay, Gould goes to great lengths to demonstrate that Pierre Teilhard de Chardin is at the very least a coconspirator. I repeat: Within the scientific community this kind of charge is the ultimate accusation. There is no more heinous professional crime than "data faking," for the social organization of knowledge depends on using shared material to build a framework of understanding. If the raw data are fabricated, the entire enterprise is in danger of collapse. An accusation of data faking, based on loosely constructed webs of evidence, is serious enough to warrant investigation. The line is thin between speculating on the identity of the hoaxer and character assassination, particularly when the object of the accusation is dead and unable to defend himself. Thus Gardiner's and Walsh's conclusions lead us to raise the question of why Stephen Jay Gould decided to do a job on Teilhard de Chardin based on circumstantial evidence, none of which is supported by the latest findings.

I present Gould's accusation in his own words:

But I do believe that a conspiracy existed at Piltdown and that, for once, the most interesting hypothesis is actually true. I believe that a man who later became one of the world's most famous theologians, a cult figure for many years after his death in 1955, knew what Dawson was doing and probably helped in no small way—the French Jesuit priest and paleontologist Pierre Teilhard de Chardin.

Note the rhetorical device of representing Teilhard as "a cult figure." While there are many serious students of Teilhard's thought, they cannot be described as a cult. At the extreme they form together in a scholarly association. Gould is not just describing Teilhard, but setting up the reader for the attack. We thus need some background information.

Teilhard de Chardin (1881–1955) was a paleontologist who belonged to the Jesuit order. His best-known work, *The Phenomenon of Man,* is a study of the evolutionary trajectory from the beginnings of the universe to the emergence of the human mind. The work points beyond the mind to the human spirit, moving from the first three parts, entirely biological, to a theologically oriented Book Four.

The work was not published in Teilhard's lifetime: his Jesuit superiors worried about heresy in his thought, and he was true to his vow of obedience.

It has been popular sport among hard-nosed reductionists to excoriate Teilhard, and Peter Medawar and Jacques Monod were leaders in this endeavor. Gould writes in this tradition, but he goes beyond his colleagues in moving from Teilhard's theories to an ad hominem argument about Teilhard the coconspirator.

Here I must lay my cards on the table. I am an admirer of Teilhard's biological writing. I do not understand his theology, and I recognize certain overly strong teleological perspectives, but I have found his ideas insightful and often ahead of their time. He has made clear that to understand biology is to comprehend evolution in the fullest sense. The biological community has not been kind to the memory of Teilhard, although the renowned evolutionary theorist Sir Julian Huxley was an exception.

Within this context I shall speculate on why Gould may have selected Teilhard for disgrace. I do not question that the professor believed what he wrote: I am, however, questioning why he chose to go public with it. Both Gould and Teilhard are Darwinian evolutionists, but with a major difference of perspective. Teilhard believed that evolution among the animalia has a direction leading to cephalization and hominization. Gould is a strong advocate of the random features of evolution and the notion that, if the tape were rewound and played again from the same beginning state, the result would be very different. It is clear that these two evolutionary theorists have opposing viewpoints. What is hard to believe is that Gould would have targeted Teilhard because of a difference of evolutionary philosophy.

Let me point to a strong similarity: both Gould and Teilhard are believers in punctuated equilibrium. Gould uses the metaphor of "a trend . . . is more like climbing a flight of stairs (punctuation and stasis) than rolling up an inclined plane." Teilhard uses the metaphor of water slowly heating up [stasis] and then reaching the boiling point at which a sharp change of state occurs [punctuation]. Indeed, the evolutionary discreteness of hominization is a key point in Teilhard's view about the origin of man. Yet Gould states, "In 1972 my colleague Niles Eldridge and I developed the theory of punctuated equilibrium." *The Phenomenon of Man,* without using the phrase, resonated with punctuated equilibrium many years earlier: "With that understood, there is nothing surprising in our finding, when we look back, that everything seems to have burst into the world *ready made.*" I have no doubt that Teilhard was committed to the concept embodied in punctuated equilibrium when Gould and Eldridge were in short pants, although they certainly are to be credited with the terminology. Yet I cannot accept the idea that Teilhard's views on punctuated equilibrium prior to Gould and Eldridge's could have been the basis of Gould's barbed arrows.

A strong difference between Teilhard and Gould is the deep religious faith of the former as contrasted to Gould's sympathies with the agnosticism of Thomas Huxley. I think that much of many scientists' distrust of Teilhard stems from the fact that he was a priest. Evolutionary biologists from Thomas Huxley forward had reason to be at odds with the clergy over evolution, but Teilhard was of course trying to make evolution a part of religion. Teilhard, according to Julian Huxley, "forced theologians to view their ideas in the new perspective of evolution, and scientists to see the spiritual implications of their knowledge." In any case, it is inconceivable that anyone with Gould's interests in fairness and civil liberties would have fingered a scientist because of his religious beliefs.

None of the possible explanations of Gould's motivation have convinced me. Although my fellow scientist in a later article wrote that he was not out to destroy Teilhard, he accused him of the ultimate scientific misdeed. His tenuously argued accusation has now been further shown to be without merit. I am left puzzled. Why did he choose this subject for his essay?

In a sense I am being unfair in directing this essay solely toward Dr. Gould. After all, everyone who improperly identified the hoaxer on the basis of circumstantial evidence must answer to their reasons for the accusations. Gould's essay is a case study in that literature. Why then do I pick this instance of finding the hoaxer? Maybe it's just that I expect a higher standard from Harvard professors.

Continuing Education

Several years ago, following a birthday celebration for Murray Gell-Mann, I wrote an essay recalling an undergraduate physics laboratory where he and I had carried out an experiment described in the text as "Mechanical Equivalent of Heat by Puluj's Method." The procedure involved turning a crank on a friction device and measuring the temperature rise in an attached calorimeter. In the essay I expressed my regret at being unable to track down any further information on Puluj, who presumably devised this classical form of the experiment. And so I remained in ignorance for many years.

Then, one day a few weeks back, the wonders of e-mail brought me a note from Roy Schmeltzer, who had read the original essay and forwarded a message from *Ethnic Newswatch* reporting on a piece in the January 8, 1995, issue of *Ukrainian Weekly* on the 150th anniversary of the birth of physicist Ivan Puluj. I was not at that time unambiguously sure that this was the individual I was seeking, but the dating, the discipline, and the rarity of the name made it extremely likely. In any case, Ivan Puluj is interesting enough in his own right that I take this opportunity to share some of my newfound information and relate my search for certainty on Puluj's method.

Ivan Puluj was born in the Ternopil region of the Ukraine on February 2, 1845. Like many Eastern European scientists of his time, he received his advanced education in one of the fast-developing German scientific centers, earning a doctorate in natural philosophy from Strassburg University in 1877. He moved to Vienna, where he served as *privat-docent* at the University.

In 1884 he was called to Prague as professor at the German Higher Technical school, a position he seems to have held until his retirement in 1916. He died in Prague on January 31, 1918.

Now that the gates have been opened, the wonders of on-line searching are producing further references, and I have come to realize Puluj's full distinction as a physicist. During the period 1880–82, he carried out a series of experiments on cathode rays. In 1889, the British Physical Society published *Physical Memoirs,* selected and translated from foreign sources. It included "The Thermodynamics of Chemical Processes" by H. von Helmholtz, "On the Conduction of Electricity in Gases" by W. Hittorf, "The Continuity of the Liquid and Gaseous States" by J. D. Van der Waals, and "Radiant Electrode Matter and the So-Called Fourth State" by J. Puluj (Puluj was called Johann, the Germanic form of the Ukrainian Ivan).

The company in which we find Puluj includes some of the most distinguished nineteenth-century European physicists.

Helmholtz, Hittorf, and Van der Waals are icons of physics and physical chemistry. Given this eminence, I am left wondering why Puluj is nowhere to be found in the *Dictionary of Scientific Biography*. The other three appear prominently in those volumes. The vagaries of historical memory are, of course, well known. In my defense, I must note that during my original search I did inquire of one of the country's leading historians of physics. He also had not heard of Puluj.

The article in the *Ukrainian Weekly* suggests that Puluj was a codiscoverer of X-rays along with Wilhelm Roentgen. Puluj, who published his finding shortly after Roentgen, referred to the radiation as "Roentgen Strahlen."

His subsequent distinguished career included many honors, election to scientific societies, and rectorship of the school in Prague from 1885 to 1890. He developed a number of patents in electrical equipment and is regarded as a pioneer in electrical engineering.

Along with his scientific achievement, Puluj was a coauthor of the first complete translation of the Bible into modern Ukrainian. He was also an active advocate of the political and cultural rights of the Ukrainian people.

The *Ukrainian Weekly* article suggests that Puluj's nonscientific activities contributed to his loss of standing as an important scientist. He was briefly mentioned in the first edition of the *Ukraine Soviet Encyclopedia* (1963), but not in the second edition (1983). The weekly suggests this was due to the KGB's opposition to the Ukrainian cultural revival and the Soviet opposition to the religious implications of the translation of the Bible.

Having come this far, my research librarian plunged ahead to find a small book by Puluj in the Harvard University Library and two libraries in Lyon, France—*Uber die Abhangigkeit der Reibung der Gase von Temperatur* was published in 1876. There is also a 1971 biography by IUrii Hryvniak entitled, *Prof. doctor Ivan Puliui, vynakhidnyk prominnia "X": profil zhyttia i diialnosti velykoho naukovtsia ta doslidnyka,* published by the Association of Ukrainians in Great Britain. The notes indicate the book is in Cyrillic and lists the title as *Prof. Ivan Puluy, discoverer of the ray "X".* Alas, one awaits an English translation.

Still, I had to admit that there was the possibility that I might not have found the right Puluj. An e-mail query went off to the convener of the 150th anniversary conference, Dr. O. Derzhko of the Institute for Condensed Matter Physics in Lviv, Ukraine. His reply: "Thank you very much for your letter concerning Ivan Puluj. I have passed it to Professor Roman Gaida. I hope he will contact you soon."

And I am happy to report I have just received from Professor Gaida a detailed paper on Puluj and his work, including the long-sought-after references to the experiment on the mechanical equivalent of heat. With gratitude to Professors Derzhko and Gaida, I present these references:

1. Johann Puluj: *Über einen Schulapparat zur Bestimmung des mechanischen Wärmeaquivalentes*. Sitzungsberichte der Mathematisch-Naturwissenschaftlichen Classe der Kaiser-lichen Akademie der Wissenschaften (shortened: Wiener Berichte). II Abt. 1875, Bd.71, S.677–686.

2. J. Puluj: *Beitrag zur Bestimmung des mechanischen Wärm-eaquivalentes*. Ibid. S.53–60.

Well, all doubts are erased. I have filled in a detail that was missing in my sophomore year of college, and in the process have become intrigued with a very interesting scientist. I think that this experience can best be described as continuing education. In these days of complexity, one sometimes feels a nostalgia for the absolute certainty and simplicity of the mechanical equivalent of heat by Puluj's method.

Saints Nicholas

O N A G R A Y, misty November day in Egmond aan Zee, I wandered through town, trying to reclaim my sense of time and space after an all-night flight across the Atlantic. The main street of this village on the Zuider Zee in the Netherlands clearly catered to tourists and was lined with small quaint shops.

I was peering into the window of an antique store when my jet-lagged reverie was interrupted by the strains of band music in the neighborhood. The sound drew me to a square a few hundred meters from the seashore, where a small brass band was holding forth with festive music. Then came a fanfare, and all eyes turned to the sea, where a tractor was pulling a boat onto a wheel bed and then up the incline to the square. On the boat was a white-bearded man in long ecclesiastical garb wearing a cross and carrying a crook. He was accompanied by eight companions in period clothing from what I would guess was the 1700s. The leader was Sint Klaes, or Saint Nicholas, and the assistants were arrayed in blackface. Given current North American sensitivities, it was quite shocking to see Saint Nicholas's companions in this particular makeup. The children in Egmond aan Zee gathered around the saint, and his assistants performed acrobatics and gave out candy to the little ones.

A local informed me that Sint Klaes Day was December 6, but the merchants of Egmond aan Zee had moved up the celebration a few weeks for a longer shopping period before December 25, the day when Santa Claus delivers gifts. The distinction between Sint Klaes and Santa Claus was sufficiently enigmatic that a modest amount of historical research was in order.

The first Saint Nicholas was Bishop of the city of Myra in the province of Lycia in Asia Minor. On a contemporary map, this is the Mediterranean city of Demre in southwestern Turkey. Although the historical record is meager, the Saint is believed to have lived in the third and fourth centuries and to have died around 350 A.D. For reasons that seem lost in history, Saint Nicholas of Myra has become a patron of sailors, captives, children, unmarried women, and pawnbrokers.

One of the enduring legends is that the saint provided three bags of gold as dowries for the daughters of an impoverished neighbor. These dowries rescued the young women from the sad fate that would have awaited them. The three bags of gold are somehow related to the three gold balls that used to hang above pawnbrokers' shops. (When one is a patron saint of several constituencies, some multiplicity of legends is required; that the number of gold objects is three cannot be without significance in the realm of Christendom.)

History does not tell us why Nicholas became such a popular

saint, but in the year 1087 his remains were taken from Myra to the city of Bari in Italy. Following the Reformation, the enduring Bishop of Myra emerged as Sint Klaes in the Netherlands, and his day, December 6, became an occasion of celebration. Why the good burghers of Holland should have related to this thirteen-hundred-year-old figure is a matter for speculation left open to future church scholarship.

With the establishing of the Dutch colony of New Amsterdam on the shores of the Hudson River, Sint Klaes crossed the Atlantic as the settlers brought their tradition with them. In the New World, Sint Klaes underwent a transition and emerged as Santa Claus. The date of his annual appearance was moved from December 6 to December 24 or 25, and his home was moved from Myra to the North Pole. By the time Santa Claus returned to the Netherlands, he was sufficiently changed that he was regarded as other than Sint Klaes, hence the Dutch folk traditions that honor Sint Klaes on December 6 and await the gifts from Santa Claus on Christmas Eve. And so my modest researches made clear to me the significance of the events I had witnessed on the shores of the Zuider Zee.

However, all true research is open-ended and my attempts to find out about Sint Klaes revealed nine other saints with the name Nicholas. I feel some obligation to share this information:

1. Nicholas of Studites. Born in Crete, died in 863 at the monastery of the Studios at Constantinople.
2. Pope Nicholas I. A Roman who served as Pope from 858 to 867, he is known to have effected the conversion of Bulgaria.
3. Nicholas the Mystic. Died in 925. He was a patriarch of

Constantinople who was deposed by Emperor Leo after he had denied the monarch permission to marry for the fourth time.

4. Nicholas of Chrysoberges. Patriarch of Constantinople from 983 to his death in 996. Constantinople thus produced three Saints Nicholas.

5. Nicholas Peregrenus. A Greek who lived a brief life from 1075 to 1094 and died in southern Italy, where he had wandered about carrying a cross and crying out "Kyrie eleison."

6. Nicholas of Tolentino, 1245–1305. An Italian Augustinian priest known for his work among the poor.

7 & 8. Nicholas Pieck and Nicholas Popple. Two native Dutchmen, members of a group martyred at the Franciscan friary at Gorkum. Modest efforts to find out about their martyrdom have to date yielded no further information.

9. Nicholas Foster. The last saint named Nicholas. He was born in Valencia in 1520 and became a Franciscan in 1537. As an itinerant preacher, he scourged himself before every sermon, and he died in 1582.

There we have it. The pipers of Egmond aan Zee drew me to a couple of libraries, an interesting conversation with a colleague, and a somewhat deepened view of hagiography. Well, I think that is what education is supposed to be.

It does leave us still wondering why the acrobats were in blackface. "Tradition," said my Dutch host.

II

Language

Biological Themes in Literature

FOR SOME YEARS I have been practicing English without a license, giving a course called "Biological Themes in Literature" in the English department at George Mason University. The course seems to be attracting some interest, so it might be appropriate to discuss this approach in more detail. A caveat is in order: there is a hidden agenda, teaching biology to English majors who might otherwise reject science and teaching literature to biology majors who might shy away from the humanities.

The class begins with a series of poems by D. H. Lawrence on the mating behavior of elephants, whales, tortoises, and fish. This is anthropomorphism in the extreme, wherein Lawrence projects his mind and phallus into the animal world. But when it comes to a fish that mates with "not one touch," Lawrence fails to empathize, concluding, "His God stands outside my God." Lawrence's powerful poetry and the topic of mating suggests to students that there may be something to this subject material.

The next work is Mary Shelley's *Frankenstein*. Written in the early 1800s, two decades before the cell theory, this powerful novel raises major questions about the definition of life, the social responsibilities of scientists, and the societal response to a creature whose "original sin" is ugliness. The work of a nineteen-year-old woman in an age of few women authors, *Frankenstein* stimulates an often passionate discussion of the lives and perspectives of writers and of the social responsibilities of scientists. If all goes well, these themes continue to evolve for the rest of the course.

Henry David Thoreau's *Walden* can be read at many levels, including a review of the ecology of Walden Pond and the surrounding habitat. As a work of natural history, the book has much to say of the flora and fauna of the Walden area and the limnology of Walden and surrounding lakes. It also speaks to environmental issues and the effects of the Industrial Revolution on American society. A homespun yet deeply philosophical document, *Walden* asks us to consider who we are, where we have come from, and whither we are going. Thoreau speaks out against the materialism of his own time, and therefore speaks even more pressingly today to the greater materialism of our time.

Moby Dick is one of the major American novels. Because it deals with whales and whaling, the work must venture into whale anatomy and physiology—but Herman Melville goes far beyond the absolutely necessary material to delve into cetacean taxonomy, ethology, and ecology. Because the course ranges wide and the semester is short, we do not read all of *Moby Dick*. Included are the chapter "Cetology," in which Melville constructs his own taxonomy of whales; chapters on the representation of whales; a chapter called "Squid," where the author tells of the giant decapod (or Kraken) and squid in the diet of sperm whales, whales as food, and the difference between baleen and toothed whales; and a chapter on fossil whales. Melville's biological erudition was impressive, and the inclusion of this material adds to the development of the novel. A major question discussed is why the novelist includes so much biological detail.

Next on the agenda is *Inherit the Wind,* the dramatic rendition of the Scopes trial by Jerome Lawrence and Robert E. Lee. The play deals with Bertram Cates (John Scopes), charged with the crime of having taught evolution in Tennessee in contravention of state law. Because the trial attracted national attention, Matthew Harrison Brady (William Jennings Bryan) is brought in for the prosecution, and Henry Drummond (Clarence Darrow) comes to Tennessee to join the defense. The work deals with evolution and creation, church and state, science and religion, and other issues that continue to resonate through American society. Through Drummond, the play expresses an understanding of the various arguments and the strong emotions aroused by these ongoing issues.

One of the interesting details in *Inherit the Wind* is that none of the townspeople have read Darwin's *Origin of Species,* the "heretical" work that is at the core of the trial. This text stays on the shelf, often cited but mostly unread, in the 1990s as well. To remedy this on a local level, the class reads the first three chapters of the *Origin*. This volume, which has been of overwhelming importance in Western intellectual history, is the only actual scientific work on the syllabus.

Brave New World by Aldous Huxley is of the genre of anti-utopian novels of the middle twentieth century. A few imaginative writers looked ahead to see where the world was taking us. Like most surviving prophetic writings, these works are gloomy. However, the raison d'etre of this variety of literature is to warn the world and prompt people to change their ways. The relation of biology is clear in this novel. The morphogenetic process in *Brave New World* has been entirely moved from the uterus to a bottle. Cloning, prebirth conditioning, and childhood training produces a rigid hierarchy of individuals who love their lack of freedom. The existential pain of life is obscured by the drug soma, and death is reduced to phosphorous recycling. Consumerism is a social goal, and promiscuity is encouraged.

The next work continues the study of Aldous Huxley. A relatively unknown essay that deserves a wider readership, the short volume entitled *Literature and Science* explores how writers and scientists "purify the language of the tribe" in quite different ways. The scientist seeks to express public knowledge with precision and invents a vocabulary and syntax "designed to express the limited meaning with which he is professionally concerned."

The literary artist, on the other hand,

takes the words of the tribe and, by a process of selection and novel arrangement, transforms them into another, purer language—a language in which it is possible to communicate unsharably private experiences, to give utterance to the ineffable, to express, directly or by implication, the diverse qualities and meanings of existence in the many universes—cosmic and cultural, inward and outward, given and symbolical—within which human beings are predestined, by their multiple amphibiousness, to live and move and have their bewildered being.

Through numerous examples Huxley explores the two purifications of language.

The next item is a series of biographies of the founders of microbiology. *Microbe Hunters* has been a very important book in the early reading of many of the noted twentieth-century microbiologists, molecular biologists, and medical researchers. Paul de Kruif's series of short biographies, so hyperbolic as to be almost caricatures, nevertheless presents the basic historical development from the microscopy of Antonie van Leeuwenhoek to the therapeutic chemistry of Paul Ehrlich. The reader comes away with substantial knowledge of infectious diseases and aspects of immunology. Copyrighted in 1926, de Kruif's book retains its interest even now, more than seventy years later. However, a problem in reading it is that racist and sexist passages surprising to today's reader often turn up. Old works reflect the societies that produced them, including attitudes that are clearly not acceptable to today's society. After discussing such issues in class, the students are prepared to explore the works for what they offer and to celebrate our present sensitivities.

Microbe Hunters serves as both a technical and attitudinal introduction to *Arrowsmith* by Sinclair Lewis. The works, one fictional biography, the other slightly biographical fiction, overlap in time, for the fictional Martin Arrowsmith entered medical school a year before the real Paul Ehrlich received his Nobel Prize for developing salversan in the treatment of syphilis.

The novel itself deals with the intellectual, scientific, and personal growth of Arrowsmith: medical student, country doctor, small town public health official, clinic employee, and researcher. In critiquing the health establishment, it is quintessential Lewis—and some of the critique seems remarkably current. There is deep science about the discovery of bacteriophages (viruses that attack bacteria) and the failed attempts to use these agents therapeutically. There is much about the social responsibility of physicians and researchers. It is also an extraordinarily good novel, as the Pulitzer and Nobel Prize committees agreed.

The relatedness of literature and biology appears in several oblique ways in John Steinbeck's *Cannery Row*. Edmund Wilson accuses Steinbeck of a reverse anthropomorphism, presenting humans as if they were animals. Thus *Cannery Row* stands at the opposite extreme from D. H. Lawrence's mating poems. Doc, the owner and sole employee of the Western Biological Laboratory, is the hero of the novel, insofar as it has a hero. In the narration of Doc's collecting trips, there is introduced a genre of descriptive biology:

Doc was collecting marine animals in the Great Tide Pool on the tip of the Peninsula. It is a fabulous place: when the tide is in, a wave-churned basin, creamy with foam, whipped by the combers that roll in from the whistling buoy on the reef. But when the tide goes out the little water world becomes quiet and lovely. The sea is very clear and the bottom becomes fantastic with hurrying, fighting, feeding, breeding animals. Crabs rush from frond to frond of the waving algae. Starfish squat over mussels and limpets, attach their million little suckers and then slowly lift with incredible power until the prey is broken from the rock. And then the starfish stomach comes out and envelops its food. Orange and speckled and fluted nudibranchs slide gracefully over the rocks, their skirts waving like the dresses of Spanish dancers. And black eels poke their heads out of crevices and wait for prey. The snapping shrimps with the trigger claws pop loudly.

The characters in this book all live on the fringe of society, yet observing them in their outsider roles has a lot to tell the reader about the human condition. Like reading Thoreau and Huxley, reading *Cannery Row* in today's world asks students to check out their value systems. Many students react favorably to this exercise. In an area rife with shopping malls, the critiques of rampant consumerism as the reason for existence deserve some careful thought. These biologically related writings seem to play this role very well.

At this juncture we turn to what I call "recycling poetry," the fascinating thought that poets have written about ecological recycling centuries and even millennia before scientists. "Compost," by Walt Whitman, is the paradigmatic recycling poem. Nine hundred years before Whitman, Omar Khayyám wrote this quatrain, as translated by Fitzgerald:

> *I sometimes think that never blows so red*
> *The Rose as where some buried Caesar bled;*
> > *the every Hyacinth the Garden wears*
> *Dropt in her Lap from some once lovely Head.*

This poem has instructive lessons about religion and society, knowability, and the problems of translation.

In future iterations of this course, we may go back yet another millennium to Lucretius, who was both poet and scientist in his epic work *De rerum natura* (On the Nature of Things). He reasoned recycling from the permanence of atoms:

> Yet again, if the matter in things had not been everlasting, everything by now would have gone back to nothing, and the things we see would be the product of rebirth out of nothing. But, since I have already shown that nothing can be created out of nothing nor any existing thing be summoned back to nothing, the atoms must be made of imperishable stuff into which everything can be resolved in the end, so that there may be a stock of matter for building the world anew.

The final novel in the syllabus is *Cat's Cradle* by Kurt Vonnegut. Although it is not based on a biological theme, the novel is rich in atomic physics and chemistry and engages the question of the habitability of the world. Both extremely funny and frighteningly downbeat, it returns us to the Frankensteinian theme of the social responsibility of scientists. Like Huxley in his gloomy prophecy, Vonnegut asks us to take a hard look at where the world is going.

Ending the course on an upbeat note, we study the essay as a literary genre in modern biology. Choosing from a wide group, I have tended to rely on *Lives of a Cell* by Lewis Thomas and a collection of my own, *Mayonnaise and the Origin of Life*. These essays explore, from a biological perspective, who we are. If the students are feeling downbeat after reading Vonnegut, I have them read my essay "Optimism as a Moral Imperative."

Deconstructing Ivory Towers

As a metaphor for academia, *ivory tower* is one of those enigmatic terms that are universally used and fully understood until one starts to think about them. *Ivory tower* is almost an oxymoron, since ivory is hardly a construction material appropriate for buildings and in any case would express an opulence uncharacteristic of academic moderation and penury.

An early use of *ivory tower* occurs in that sensuous Old Testament love poem that confounds scholars of theology, the Song of Songs.

> *Thy two breasts are like two young roes that are twins.*
> *Thy neck is as a tower of ivory:*
> *thine eyes like the fishpools in Heshbon, by the gate of*
>   *Bath-rabbim.*

Ivory, particularly from elephants' tusks, was well known in biblical times, and the Hebrew *shen* and *shenhabbim* (tooth and elephant tooth) appear frequently in the Old Testament. Ivory was very costly in the past, and in the East it was usually used for inlaid work. Ahab built for himself an "ivory house," the halls and chandeliers of which were enriched with inlaid ivory. Ezekiel mentions the rich ivory ornamentation of the decks of the Phoenician ships. The prophet Amos condemns those who luxuriate on beds of ivory. King Solomon sat on a throne of ivory.

Elephant tusks came largely from Africa and India. By Homeric times, they had also found their way to Greece and the palace of Menelaus.

Son of Nestor, delight of my heart, mark the flashing
of bronze through the echoing halls, and the flashing
of gold and of amber and of silver and of ivory.
[The *Odyssey*]

So ivory was well known, and the unknown troubadour may be permitted poetic license in describing the long, ivory-hued neck of the beloved. But what has this to do with the groves of academe?

Researching this puzzle led me first to the Song of Songs in the Anchor Bible Series, a canonical work of biblical scholarship by Marvin H. Pope. (This was especially rewarding, as Professor Emeritus Pope and I were colleagues in the Pierson College Fellowship at Yale, and reading his name evoked fond memories.) Pope footnotes the modern usage of *ivory tower*, which is universally credited to the French writer Charles Augustin Sainte-Beuve. In October 1837, Sainte-Beuve wrote to M. Villemain:

Hugo, stern partisan; fought under armor,
and held his banner high in the midst of tumult:
He holds it still; and Vigny, more secret
As if in his tower of ivory, retired before noon.

There seems little question that the modern usage of *ivory tower* for intellectual and academic isolation from the rest of the world stems from this *tour d'ivoire* of Sainte-Beuve.

Since Sainte-Beuve, the ivory tower has found its way into English and various European languages, and there are many entries in volumes such as the *Oxford English Dictionary*. It is now a universal synonym for academia. Nevertheless, it seems very strange that Sainte-Beuve would have leapt over 2,200 years of usage to go from the neck of the beloved to the intellectual isolation of Vigny.

I could not quite leave the subject in so unsatisfactory a state, so I sought sources that might have influenced Sainte-Beuve. This literary savant lived from 1804 to 1869. He was a scholar of French literature from the Renaissance to the nineteenth century, so if there were early references to *le tour d'ivoire* he would undoubtedly have come across them. The only recourse seemed to be to read the collected works of Sainte-Beuve, but I am inhibited by considerations of time and a very fragmentary reading knowledge of French.

A French colleague suggested that I seek out *Trésor de la langue française: dictionnaire de la langue du XIX et du XX siècle* (1789–1960). This multivolume work is the French version of the *Oxford English Dictionary* and was compiled by Paul Imbs.

I, of course, rushed to the library and was met by the following frustration: This multivolume work was still in press and only volumes up to "Sa" were available. Thus, *ivoire* referred me to *tour,* but that volume was some time away!

So the lacuna persists, perhaps not as one of the great intellectual problems of our time. Nevertheless, I am perplexed, and when the appropriate volume of *Trésor de la langue française* arrives I shall be waiting in line, hoping that the mystery will be clarified.

When I was about to shelve this matter temporarily, a phone call from a friend with whom I had discussed ivory towers alerted me to a reference in the *Litany of Loreto*. These writings, which may date back to 1200, include in their various praises of Mary, mother of Jesus, a comparison to a tower of ivory. This opens up a path of research to be pursued at some future date.

Meanwhile, I make a special plea to readers with knowledge of the uses of *ivory tower* or *tour d'ivoire* or *migdal shen* or any other linguistic version between the redaction of the Old Testament and 1837: Kindly inform me so that knowledge in this area may be advanced.

And to those of you who would criticize this activity as being totally ivory towerish, I would reply with a quiet mea culpa. It is part of my professorial task to seek knowledge of the world, wherever that search might lead.

Reductionism Is Not a Dirty Word

I HAVE in recent years come to use the word *reductionism* with increasing frequency, as it seems to embody concepts that are difficult to find in other nouns. Of late, I have found editors, readers, and colleagues becoming a bit uncomfortable about the word, so some thought seemed to be in order, beginning with the definition in *Merriam Webster's Collegiate Dictionary,* tenth edition, 1993:

> 1 : the attempt to explain all biological processes by the same explanations (as by physical laws) that chemists and physicists use to interpret inanimate matter; *also* : the theory that complete reductionism is possible 2 : a procedure or theory that reduces complex data or phenomena to simple terms; *esp* : OVERSIMPLIFICATION.

I was surprised to find the definitions (1) so biological and (2) so confined to physics and chemistry as the ultimate ground of explanation. I find it intellectually more satisfying to consider a general use of the term, that is, one that would apply to any subject that can be arranged hierarchically to designate reductionism as the explanation of phenomena at one level in terms of understanding at the next lower level. This has certainly been applicable to biology, in which the levels are conventionally taken to be ecosystems, biomes, species, individual organisms, organs, tissues, organelles, macromolecules, small molecules, atoms, and nuclei, electrons, and other elementary particles. Of course, the lowest five levels are not restricted to biology, so the basic ideas of reductionism are somewhat broader than the dictionary view.

The bottom and top of any hierarchy of this type are going to be the most elementary particles and the universe. It is interesting that most results of research in elementary particle physics and in cosmology are mutually validating and thus introduce a curious circularity into the business of hierarchies. However, most of the questions in contemporary biology are confined to the range from atomic physics to global ecology and thus escape the circularity just discussed.

It is my sense that when the history of twentieth-century science is written, reductionism will emerge as one of the strongest themes. This will be particularly true of biology and medicine, where physics and molecular biochemistry tend to dominate the understanding of both normal and pathologic processes. This tendency toward physical explanation is not new; in the 1800s, it characterized much of research in biology and physiology, as can be seen in the work of such investigators as Claude Bernard.

Reductionism seems to have picked up steam in all areas of biology in the post–World War II era. The fact that bacterial, viral, genetic, physiologic, and biochemical studies gave great insight into mammalian and human biology was a very persuasive argument for the power of reductionism.

Why then does the word evoke such strange responses? I think it is in part because we have carried everything to its atomistic core, and there is a vague uneasy feeling that biology and humanness have somehow slipped through our hands. In an effort to avoid teleologic reasoning and vitalism, we have turned our backs on all forms of holistic and integrative thinking. We know quite well what atoms do, but we do not have as much understanding of what organ systems, organisms, and ecosystems do. We certainly lack insight in dealing with societies and economies: Knowing the detailed molecular changes in a mutation does not enlighten us as to whether the organism will be more or less fit in a particular ecosystem.

But not to worry. There is a path out of this predicament by way of complexity theory. In one area of complexity research, the theorist starts with simple entities operating under simple algorithms and explores what happens when a collection, network, or population of these entities unfolds over time. In general, possible outcomes are combinatorial, and the space they occupy becomes both very highly dimensional and dense with representative points. In some cases, fitness filters or other pruning rules may be applied to keep the evolving program from becoming explosive. Indeed, the evolving behavior may be adaptive, leading to new insights. There may in fact be emergent properties.

In modeling toward complexity, I would argue that it is very important to maintain all the constraints inherent in the reductionist program. Whatever weaknesses it may have, for the past two hundred years reductionism has been our surest guide to understanding the world and to complement that understanding with unprecedented technological development. To opt for holism without the reductionist roots is to move into an "anything goes" domain, an epistemologic approach that is open to wild irrationality.

What is so powerful about computer technology is that speed and memory are sufficient to explore the building up of models from known and understood roots that have been developed by analysis of wholes into their parts. One should not be too casual about this. Synthesis may be far more difficult than analysis, and we are just at the beginning of the computer revolution in terms of thought processes.

Nevertheless, the good news is that one can be both a reductionist and a holist. It takes self-discipline and a healthy respect for the experiments and data that led to an analytic understanding. One is then free to use the full power of modern computational science to explore a wide range of problems that are both intellectually intriguing and of great practical importance.

It is also clear that my use of *reductionism* as a simple methodologic tool devoid of philosophical meaning is not universal. For example, Pope John Paul II in his "Message for the World Day of Peace 1990" refers to "an unnatural and reductionist vision which at times leads to genuine contempt for man." The explanation of phenomena in terms of a lower level is not an assertion that this is a complete explanation, as the papal message seemed to imply.

Attributing completeness to reductionism is as unsatisfactory to the scientist as to the theologian. It is true that they may respond in different ways, but the problem is not reductionism per se but how we use science to help formulate a holistic view of humanity.

How Mouse
Became a Verb

A FEW WEEKS ago, I walked into a colleague's office and asked, "Do you know X's telephone number?" He was sitting at his computer and looked up. "Just a second. I'll mouse it up for you," he replied. A few deft moves of the hand, and shortly the desired phone number appeared on the screen. *Mouse* was being used as a verb, and I hardly noticed the difference.

Actually, I should have been attuned to the usage of the word *mouse,* since I have recently been surveying the biology literature to generate a database recording the number of times various taxa occur as experimental organisms. I retrieve the information electronically from the on-line abstracts and titles on the basis of genus name. This method works for *Escherichia, Drosophila,* and the like, but fails for common laboratory animals like *Mus* and

Rattus. Most investigators do not feel the necessity of using the Linnaean name for commonly employed laboratory mammals and birds. Thus, whereas *dog* occurred eighteen hundred times in my search, *Canis* occurred only thirty-four times. To make full use of bibliographic databases, we are going to have to start developing some conventions about taxonomic names. The important issue of optimizing the literature of science for on-line searching and on-line use by expert systems deserves more attention than it is getting.

In any case, finding a new usage for *mouse* suggested a whole series of examples of the verbalization of the names of common animals. Occasionally, these names are converted into adjectival form.

As a result, we have "he dogged my footsteps" (verbal usage) and "dog days of summer" (adjectival usage). *Cat* occurs less familiarly in the verb form, meaning to flog with a cat-o'-nine-tails. The only adjectival usage I can think of is *cathouse*.

Horse has worked its way into verbhood in such usages as "horse around." The adjective *horse* appears with great frequency. I remember a professor who used to define *horse sense* as "that good sense that horses have that keeps them from betting on people."

Cow has become a verb meaning to frighten with threats. One thinks of a line from Ezra Pound: "I ha' seen him cow a thousand men." *Chicken* achieves verbal status in the phrase "to chicken out" and a related adjectival meaning in "he's chicken." *Parrot* becomes a verb when we parrot someone else's words.

One of the interesting cases is to *rat,* meaning to turn informer. In looking for an adjective from this rodent's name, I did find *rat fink* and recalled my late friend Alex, whose comments on the history of science were enlivened by his vigorous language, including such remarks as "Volta was a rat fink!"

Pidgin English, which by and large has not yet made its way into dictionaries, uses *cockroach* as a verb meaning to steal. Thus, on Maui one might hear a phrase like "Hey, Bruddah, who cockroach my avocado?"

All of these animal verbs set me to thinking about *mouse* again, and I found to my great surprise a history of the verb form of that rodent's name going back to 1250. The earliest usage I could find means to catch mice, as in "an owl mousing in the long grass." In a related meaning going back several hundred years, to *mouse* means to handle as a cat does a mouse, as in "death mousing the flesh of men." Another meaning is to ransack or pillage, as "the beast moused my bread box." Thus, *mouse* becomes a verb by way of the rodent's being subject or object in the process being analogized.

Another series of meanings derives from the idea of searching for something. Thus, I was mousing about in the bookstores, looking for classics, and came across a very old edition of Shakespeare, which I moused over avidly.

Lastly, there is a nautical usage whereby one mouses a hook by fastening a small cord from the tip to the back. As I finally came to grips with seven hundred years of usage, an awareness emerged of the millennia of very close relationship between mice and men.

Of course, both mice and humans are mammals, and this

shared taxonomic relation provides a certain closeness. Many strains of mice have evolved to share our habitats, adapt to our dietary habits, and otherwise inconspicuously reside in our homes while we buy the food and pay the heating bills. This leads to an antipathy between homemakers and scroungers that impels the former to an attitude of "Kill! Kill! Kill!" using mechanical death traps and poisons. There seems to be no Geneva Convention in the battle between human and rodent.

M*any years* ago, the approach of winter led many deer mice into sharing our house. The family was less than pleased. My response was to purchase Havahart humane animal traps. Each morning I would walk out far into the woods and release my prisoners. One day it was very cold and snowy. When I opened the trap, the occupant got out, looked around, and nuzzled up to my shoe. I apologized and walked away.

Some fifty years ago, a major change in the human perception of mice occurred with the creation of Mickey and Minnie Mouse by Walt Disney. These two characters have such charm that it is difficult to associate them with the enemy against whom we conduct chemical warfare. This is part of a general tendency. We anthropomorphize animals and then have to deal with the schizophrenia of our interaction with the myths and our response to the creatures in nature. This is often confusing to children.

The notion of electronic mice goes back to small machines designed to get through mazes. This was one of the early experimental approaches to artificial intelligence, deriving from classical psychology experiments with rodents in mazes. There have been competitions in which the winning mechanical mouse gets through the maze in the shortest time. When the device to move a cursor emerged as something small and confined to move on the surface of the table, I suppose that *mouse* was a reasonable analogue. And so into the modern world.

III

Science

·

Thermal Underthoughts

Doing the laundry is one of those automated tasks that hardly ever inspires analysis of the scientific foundations of the process. Today, however, while washing and drying clothes I began to think about how deeply the processes being carried out are grounded in thermal physics and about certain aspects of J. Willard Gibbs's monumental work on the equilibrium of heterogeneous substances.

The process begins, of course, with the selection of load size and protocols for washing and rinsing. The parameters are volume, temperature, and time—one extensive and one intensive thermodynamic variable, and one kinetic parameter. Remember, washing clothes is a nonequilibrium process; all parameters must be considered with that caveat in mind, although few of us take that point of view.

Next, the detergent is added. It is a mixture of several components, including an amphiphile to solubilize oily substances and a chelator to solubilize divalent cations and block the precipitation of salts of such ions. The amphiphiles and oils form a coacervate that transfers the oily dirt from the clothes to the water, which subsequently carries the dirt down the drain. Rate and solubility are temperature-dependent, and some phase changes of the oily materials are likely to occur at the temperatures chosen for washing—298°K to 345°K.

When the washer is started, water enters to the predetermined level and the agitator thoroughly mixes detergent, water, and clothes. This mixing is important, because the process of diffusion is slow over macroscopic dimensions. Stirring mixes the detergent rapidly into the water, where it is solubilized. Stirring also brings fresh solution to the dirty surface and takes away soil—by dissolving it, solubilizing it in another phase, chelating it, and mechanically removing it. A well-stirred reaction vessel is a classical concept of chemical kinetics and is particularly important in a multiphase system in which diffusion is severely limited, such as a pile of dirty clothes.

Before and after the rinse cycle, water is mechanically removed from the clothes by centrifuging them against the wall of the washer and thus effectively squeezing water from them. One way to measure the fraction of dirt removed is to subtract the volume of the residual water in the clothes after the first spin from the volume of the wash water and then divide by the volume of the wash water. The operations involved are solubilization and dilution, which are ordinary processes of chemistry.

The final spin prepares the clothes for drying, and here again, it is important to leave a minimum amount of water in the system to conserve energy. In the dryer, water is removed by converting it from the liquid to the vapor phase. The energy required is very substantial; thus, thermally drying clothes is slow and expensive. It is slow because the temperature must be kept relatively low to avoid damage to the clothes, and it is expensive because the enthalpy of vaporization is 407 kilojoules per mole (the equivalent of about 540 kilocalories per gram).

When the clothes are moved to the dryer to begin the thermal-phase transition just discussed, it is necessary to clean the lint trap, an act I dutifully perform. Of late I have taken an interest in the fluffy, semicoherent mass that is removed. I have upon a few occasions determined that the weight of this lint ranges from 0.5 to 1.0 gm for each 5-kg load of dry, dirty laundry. The existence of lint forces me to confront one of the deepest principles of thermal physics, the Second Law of Thermodynamics.

Examined under a dissecting microscope, the contents of the lint trap can be seen to consist of a mat of short fibers. There is clearly a mixture of fiber types, which must be breakdown products from clothes and towels and sheets, along with lint these items have picked up from the carpet, furniture, and other textiles. All of this seems unremarkable, since almost everything that is subjected to this cleaning cycle is made of fibers of wool, cotton, linen, or synthetic materials. However, you won't find much linen in my lint trap.

Woven fibers have been used for human clothing going back to earliest civilization. (The 5,300-year-old "Ice Man" was wearing a cape woven of grasses when he was recently discovered in the Alps.) But woven fibrous materials are structures that are far from equilibrium; they are relentlessly subject to decay, according to the Second Law of Thermodynamics. This decay may be chemical (for example, hydrolysis of the fiber polymers, oxidation of the reduced structures) or mechanical (for example, ripping, unwinding, the breakdown of fibers into fibrils). Mechanical decay produces lint. The chemical processes move toward the thermodynamic ground state by weakening the fibers and accelerating the breakdown. Lint bears witness to the first step in the breakdown of my underwear under the inevitable attack of the Second Law.

Since the loss per 5-kg load of dirty laundry is about 0.75 grams, I can formulate the following equation:

$$F = e^{-an}$$

F is the fraction of cloth remaining, n is the number of times I wash an item, and a is the decay constant. From the data given, when $n=1$, $a=1.5 \times 10^{-4}$.

Assuming that by the time the clothes have lost 1.5 percent of their fiber they are no longer fit to wear, they can be expected to withstand one hundred washings, so the problem of decay is of no immediate practical importance. Of course, if washing removes as much fiber as drying, that figure drops to fifty washings, and if wearing the clothes takes an equal toll, they will survive only thirty-three launderings. Without making other detailed measurements in addition to weighing the lint, it's not going to be easy to pin down these numerical estimates.

Perhaps more important than the practical aspect is the theoretical one. I am being treated to a vivid, everyday example of the Second Law of Thermodynamics—the fact that systems that are far from equilibrium break down and degrade. There is no way to escape this inexorable law of nature except to supply energy to perform work to reverse the degradation. The human body does this quite well: Constant molecular synthesis counters molecular breakdown. That's what basal metabolism is all about. But clothing manufacturers have not yet managed this process. In any case, clothes that last forever would be a capitalist mistake, as was demonstrated in that wonderful movie starring Alec Guinness, *The Man in the White Suit*.

Well, my laundry is done, and I have no thermal underthoughts about folding it and putting it away. So I had best excuse myself and get on with it.

The Secret of Life

THE PHYSICAL BIOSCIENTISTS of my generation have been much influenced by the essay "What Is Life," by physicist Erwin Schrödinger, one of the founders of quantum mechanics. The enigmatic nature of life looms, somehow, behind most biologically directed research and is referred to as "the secret of life" by naturalist Loren Eiseley. He explores the autumn woods and concludes that the ultimate questions will not be answered but are, in Thomas Hardy's words, "but one mask of many worn by the Great Face behind." Eiseley expresses his unease with reductionism as follows: "I have come to suspect that this long descent down the ladder of life . . . will not lead us to the final secret. In fact I have ceased to believe in the final brew or the ultimate chemical."

I too have roamed the fall woods, collected specimens, followed whales in small boats, and worked in the laboratory, but I do not share Eiseley's distrust of the "descent down the ladder of life." For the descent does not lead to one ultimate chemical, but to a remarkable collection of chemicals forming that great network of living chemistry which adorns my wall, the exquisite chart of metabolic pathways. What makes the chart so deeply meaningful is that every one of the living creatures on which Eiseley focuses has a set of metabolic reactions that map directly onto it. Putting aside the diversity, this is the unity of all life.

Because this metabolism is common to everything, it must reflect or fully duplicate the chemical reaction pathway of the universal ancestor four billion years ago. Modern biochemical information obtained in today's laboratories thus allows us to look back into the past, even into the dawn of primordial life on earth. For the biochemist, this chart is a Book of Genesis. The act of peering at it on the wall inspires an awe about the secrets of life. The answer must somehow be extractable from the few hundred chemicals and connections between chemicals that are arrayed before me. If my search for the solution to the mystery is less romantic than Eiseley's, it is nonetheless deep and strongly felt.

In musing on the chart, I have always felt that at the core of the reaction network there must be a special significance to the tricarboxylic acid cycle (TCA), which is familiarly designated the Krebs cycle after Hans Krebs and is also known as the citric acid cycle. When I learned of this remarkable cycle many years ago, it was the center of bioenergetics, the pathway through which acetate passed on its way to carbon dioxide and highly reduced cell components. And so I had long thought of it principally in terms of bioenergetics, until I began thinking about what metabolism has to tell us about the origin of life.

All living organisms are either autotrophs or heterotrophs.

Autotrophs synthesize all their molecular constituents from carbon dioxide and inorganic compounds of nitrogen, sulfur, phosphorus, and trace minerals. Heterotrophs require some organic compounds derived from autotrophs. Heterotrophs cannot exist without organisms that fix carbon dioxide, and therefore their complete metabolism relies on some pathways in the autotrophs. When one focuses on the incorporation of carbon, a new role of the TCA cycle becomes apparent: The metabolic pathways leading to sugars, fats, and amino acids all originate in this network of reactions and in a few neighboring reaction pathways. The cycle is not only the energy core, it is the engine of synthesis for the major monomers of all biomass. It follows that the first biochemically complete organisms would have, of necessity, possessed this cycle.

In the 1970s and 1980s, the discovery of a new group of microorganisms demanded a total revision of our perception of the TCA cycle. These species, of which *Hydrogenobacter thermophilus*, *Chlorobium limicola*, and *Desulfobacter hydrogenophilus* are examples, possess a reductive TCA cycle. They run the loop backward from the direction followed by oxidizers. In a reducing atmosphere, they incorporate carbon dioxide and hydrogen as reducing equivalents and produce two citrates for each initial molecule of citrate. This is a carbon-fixing, self-replicating cycle for organisms growing in reducing environments such as those found in deep oceanic trenches, swamps, and certain muds.

Without that picture of the reaction pathways—which is worth a thousand words—I must repeat what these organisms achieve: (1) They incorporate carbon dioxide and inorganic nitrogen, phosphorous, and sulfur. (2) They make all the biochemicals used in core metabolism. (3) They get their energy from combining reductants, such as hydrogen, in an environment with oxidants, such as sulfur. (4) At the core of their metabolism, they have a two-loop network of eleven compounds that is chemically self-replicating and contains the precursors of all biosynthesis. (5) These core compounds are common to the metabolism of all organisms.

Now suppose for a moment that I could go into the laboratory and get the reductive TCA cycle to run without enzymes. You might conclude that this was the framework of the earliest biochemistry and that we had found a self-replicating chemistry that could lead to life. You might conclude that we had found the secret of life.

Well, we are far from coming to closure on this one. Many days will be spent with smelly beakers, dirty test tubes, and clogged chromatography columns. In the end it will lack the beauty of Eiseley's great outdoors, but we might have descended the ladder to the final secret that the naturalist would have denied to us. I find the metabolic pathways chart no less awesome than the fall woods. They are but different masks worn by "the Great Face behind."

Immortality

A RECENT VISIT to Antelope Island in the Great Salt Lake of Utah as well as a visit to Temple Square in Salt Lake City prompted me to think about some experiments Arthur Skoultchi and I performed thirty years ago. Their significance is still puzzling.

Great Salt Lake is the home of species of the genus *Artemia*, brine shrimp able to live in the extremely high salinity of this and various other similarly salty bodies of water. In season, *Artemia* females lay eggs by the shore in areas that dry up, leaving large deposits of dehydrated eggs, small spheres a millimeter or less in diameter. When the water rises annually, the eggs are rehydrated with brine, and in about a day small shrimp emerge and begin swimming about.

We had asked the question Would complex living organisms survive temperatures near absolute zero? and *Artemia* seemed like the experimental organism of choice. It is a eukaryote, a crustacean, a brachiopod. The egg is more than a fertilized ovum; it has undergone a number of morphogenetic stages on the way toward being a free-swimming nauplius. For this experiment, the dehydrated state meant it would not be necessary to contend with ice-crystal formation, which seems to destroy most cells of eukaryotes on freezing.

The experiments were very simple. We hatched the eggs on pieces of filter paper soaking in a concentrated salt solution. We counted the total number of eggs and the number of eggs that hatched. The ratio of these values (the hatch rate) was more than 50 percent. We put dried eggs in capillary tubes and sealed the tubes. One group of tubes was kept at room temperature and a second was cooled to 2° absolute. After six days, the latter group was rewarmed. The hatch rate was more than 50 percent for each group. The experiment was not unique. A number of other investigators have reported survival of organisms at temperatures ranging from 5°K for liquid helium to 78°K for liquid nitrogen. I suspect that our experiment used the lowest temperature for the longest time for the most complex biological entity investigated. (A. I. Skoultchi and H. J. Morowitz, "Information Storage and Survival at Temperatures Near Absolute Zero," *Yale Journal of Biological Medicine* 37:158, 1964.)

Consider for a moment what things are like at 2°K. All molecular motion has ceased. The chance of any chemical bond being in its first upper vibrational state is negligible. Any structure held at this temperature will remain unchanged forever. Suppose it is a living organism such as an *Artemia* egg. It too retains its structure forever if kept at 2°K; it is unchanging and immortal. When warmed and placed in brine solution, it will hatch. There seems no other interpretation of the experimental data.

Two questions arise from this kind of analysis: What is the meaning of survival near absolute zero, and what are the implications for long-term human survival? The second of these is easier to get at. We do not know how to freeze a living human and then revive him or her, because of ice-crystal cell death. We can, however, store some mammalian cells by suspending them in solutions of glycerol or dimethyl sulfoxide and freezing very quickly, which inhibits ice-crystal formation. Thus, bull sperm as well as other mammalian cells are routinely stored at liquid nitrogen temperatures. There seems no reason whatsoever that these cells could not be taken to 2°K, where they would survive forever. Thus, we could have immortal bull sperm starting even now, although immortal bulls are a technological and perhaps a more fundamental problem currently beyond our grasp. Will these problems be solved for humans, making immortality possible? There is no way of knowing, but given the reported experiments, it would not seem to violate any law of science at the molecular level.

It must be pointed out that life at 2°K is uninteresting in the extreme. Nothing happens, and I do mean nothing. That may be the trade-off. The longer you live, the less happens—hardly a description of Shangri-la.

Let's return to the meaning of survival of complex eukaryotic life at or near absolute zero. It implies that what we call life requires continuity of structure but not of process. This comes as a surprise, because we tend to think of life as a curious combination of structure and process or, as we more conventionally say in biology, a combination of structure and function. These cryogenic experiments imply that at the cellular and even at a certain multicellular level, the structure, when released by random thermal energy and water to do its thing, will go ahead and carry out all necessary functions. The essential information of life is structural, and if the structure is preserved, function will resume when conditions are right.

What are the most complex adult animals that appear to be candidates at present for cryogenic immortality? Some species of rotifers, nematodes, and tardigrades will survive liquid nitrogen and therefore will probably survive and persist at 2°K. These animals have elaborate nervous systems of several hundred cells as well as other organs for specific physiologic functions. They are less than a millimeter long, and small size may be a prerequisite for low-temperature survival. Human blastulas are also stored at liquid nitrogen temperature, which implies a kind of human immortality, at least in a genetic sense. These simple experiments once again call on us to give some thought to who we are and what we are.

eturn now to Salt Lake City. At Temple Square, I heard some talk of immortality. Indeed, the large genealogical databases maintained by the Church of Jesus Christ of Latter-Day Saints stem from beliefs about a kind of immortality that goes, I believe, far beyond the physics and biology that we have been discussing.

Well, my trip to Utah led me to subjects I hadn't planned

to engage. Between the metaphysical immortality of Temple Square and the physical immortality of Antelope Island, there's a lot to ponder. Now, where do I travel to find the fountain of youth, so that I have enough time to think through all of these difficult thoughts?

Proctological Truth

T HE History of Medicine Library in the Sterling Hall of Medicine at Yale University is a room that seems to appear out of the nineteenth century, or perhaps the eighteenth. Over the fireplace hangs a portrait of Andreas Vesalius attributed to Jan van Calcar, a student of Titian's, who is believed by many historians to have been the artist of the wonderful drawings in Vesalius's landmark book of anatomy, *De humani corporis fabrica*. High above the portrait hangs the Vesalius family's coat of arms, three weasels in vertical display. In one corner is a portrait of John Farquhar Fulton, Keeper of the Medical History Collections from 1949 to 1960, and in another, a bust of Harvey Cushing, who directed the study of the History of Medicine from 1937 to 1939.

This is truly a tranquil portion of the universe, a mecca for bibliophiles. There are seldom more than a few patrons, and they are a quiet lot. Occasionally, a user seems to be resting his eyes, reevaluating, or perhaps even sleeping. But who knows? I have on occasion so closed my eyes and reevaluated in this room.

Sometimes, when I have a few moments to spare, I browse the shelves and am rewarded by intriguing titles. Consider, then, my surprise when I encountered a book spine emblazoned in gold letters *The Romance of Proctology* by Charles Elton Blanchard (the AMS Press 1978 reprint of the 1938 Medical Success Press edition). Since no unannotated pornography or prurient writings enter the open shelves of this sanctum, I had to conclude this book was the work of an individual whose enthusiasm for proctology knew no ends.

Yet the very semantics of the romance of proctology lends itself to a range of offbeat humor, as we all know from the scatological jokes that circulate even in a great medical school. When the author of the work before me discusses the writings of the great early rectal anatomist Giovanni Battista Morgagni (1682–1771), he translates the title *De sedibus et causis morborum per anatoment indagatis* as *On the Seats and Causes of Diseases* and compares its length to that of *Gone With the Wind*. Gone with the wind, indeed. Blanchard also refers to the rectum as "the posterior gate of the human citadel."

Much of the historical portion of the romance deals with the preproctology era before the nineteenth century. The modern epic begins with John Hilton (1804–78), anatomist at Guy's Hospital, London, who achieved a kind of immortality in describing Hilton's line as "that obscure landmark in the anal canal." The next advance in proctology was the founding of St. Mark's Hospital at Bath in 1835 and the work of Frederick Salmon, who named the institution "The Infirmary for the Relief of the Poor Afflicted with Fistula and Other Diseases of the Rectum." Salmon established rectal surgery as a specialty at a time when surgery was just becoming a major branch of medicine. Salmon was followed at St. Mark's by William Allingham.

I suppose it is only my perverse sense of humor that leads me to chuckle over Dr. Blanchard's turn of phrase: "We shall give attention to Allingham's teaching because it was at his feet that several American proctologists sat to learn rectal surgery." What do we do with that metaphor, or with the following strange sentence: "These pioneers were earnest searchers after proctological truth"?

The romance continues with chapter 5, "America Takes Up Proctology," which begins with Joseph M. Mathews of Louisville, Kentucky, author of the 1893 *Treatise on Diseases of the Rectum, Anus, and Sigmoid Flexure* and first president of the American Proctologic Society in 1899.

The chapter then goes back to the work of Milton W. Mitchell and the disputes that accompanied his use of injectable phenol in olive oil as a nonsurgical treatment for hemorrhoids. Equally disputed was his practice of selling the "secret remedy" and selling territorial rights for other physicians to acquire the medication. Mitchell practiced throughout the Midwest—moving from Louisville to Concord, Illinois, from where he also practiced in Jacksonville, Illinois, and St. Louis. He then moved to Carrollton, Missouri, and subsequently traveled on to Wichita and Coldwater, Kansas. One has the sense that he treated every hemorrhoid in the central United States. In any case, a battle raged between establishment proctological surgeons and Mitchell's followers, who did "ambulant proctological work." The intensity of feelings is seen in this quotation from Dr. Allingham:

> I have read in American pamphlets that the injection of carbolic acid into internal piles is very commonly practiced there in America; that "shoals of quacks" perambulate the country, armed with a hypodermic syringe and a bottle containing a so-called secret remedy; this remedy being carbolic acid diluted in different ways and of different strengths.

Proctology, as a discipline, seems to have had a hyperbolic rhetorical aspect in America during this period. In 1886, Dr. W. E. Ryan of Indiana published a book entitled *Aphorisms in Rectal Disease*. In 1891, Dr. E. H. Pratt of Illinois published *Orificial Surgery*.

At this point in *The Romance of Proctology,* Dr. Blanchard's true agenda emerges. As a follower of Dr. Albright, a pioneer in nonsurgical approaches, he became committed to the wisdom of using ambulant methods in treating rectal diseases, and the romance deals with the acceptance of these methods. The book was written in 1938, and "the modern period" he deals with is post–World War I.

The last six chapters, while largely clinical in nature, are remarkably emotional, in contrast to current writings about medical procedures. Chapter 7, "The Story of a Personal Transition," presents one doctor's choice of the ambulant approach in language appropriate to a religious conversion. To his mentors, he offers "praise and appreciation with the circle of my soul rather than the words of my mouth." Chapter 8, "An Office Episode," uses a Galilean type of dialogue to argue the case. The book ends on a terribly serious note, followed by fourteen pages of drawings of proctological instruments.

I grow drowsy in the library and think of taking this book out on loan. I notice that no one has so disturbed it since 1980. I have certainly acquired a knowledge of the history of proctology that goes way beyond what cultural literacy would require of me. It hardly seems like trivia suitable for cocktail party conversation. But with the faith that no knowledge is useless, I will one day find an opportunity to make an assertion about proctological truth, whatever that turns out to be.

Heaven, Hell, and Bioenergetics

OUR IMPROVED UNDERSTANDING of ecology and bio-chemistry has made us acutely aware of the dominant role of energy in all living processes. Bioenergy is what we buy at the supermarket and at the gas station, and it is the flow of this energy that brings order and organization to the world of living things, great and small.

For many years, it was believed that the sole source of energy for all biota was sunlight falling on plants and algae, thereby powering the photosynthetic production of all the world's food. The sine qua non of the biological food chains was a gift from the heavens that enriched the earth. Later, it was realized that the living world requires a constant flow-through of energy that incorporates a sink as well as a source. The sink was found in the cold of outer space. The planet radiates infrared to the universe to balance the receipt of radiation of shorter wavelengths from the sun.

Then a revolution in geologic thought dealing with the motion of tectonic plates led to the discovery of a second energy source for ecosystems. First, consider the setting. There are regions of the ocean floor where newly formed earth crust rises into place and, as a result, tectonic plates slowly move apart. When experimental submarines capable of descending to great depths explored these midocean ridges, to everyone's great surprise, a wide variety of unusual animal forms was found. But the ridges are thousands of feet below the photic zone, and thus are regions of absolute darkness. How can life exist there without energy from the sun?

The investigation of that question led to an understanding of the second energy source. From the tectonic ridges, heated by such subcrustal phenomena as radioactive decay and gravitational condensation, there emerge very hot compounds of iron and sulfur in reduced form, from the perspective of oxidation-reduction chemistry. Highly specialized bacteria are able to combine these reduced compounds with oxidants in the surrounding ocean, yielding sufficient energy for the bacteria to incorporate carbon dioxide and grow. The bacteria are eaten by tiny animals, some of which are eaten by larger animals, and so on, leading to an entire ecosystem powered by planetary oxidation-reduction energy, in contrast to surface ecosystems that are powered by solar photons.

Thus, while one form of energy descends from the heavens, the second bubbles up from the hot subterranean depths accompanied by superheated sulfur-containing gases. It is as close as one can imagine to Dante's *paradiso* and *inferno*.

At a casual glance, the two energy sources seem very different, but at the chemical level they are closely related. The first step in photosynthesis is splitting of the water molecule into completely oxidized $\frac{1}{2}O_2$ and completely reduced 2H. The oxygen is given off as an atmospheric gas, and hydrogen appears in the cell as reduced compounds of the form AH_2. The subsequent reactions largely involve the synthesis and hydrolysis of adenosine triphosphate and ultimately lead to the recombination of 2H and $\frac{1}{2}O_2$ to form water. The chemosynthetic cells use oxidation-reduction reactions of environmental molecules to produce reductants that also are ultimately oxidized, leading to the production of adenosine triphosphate. The network of intermediary reactions is the same, independent of the primary energy source.

This similarity of the biochemical networks leads us to the origin of life. Which has precedence, heaven or hell? Did photosynthetic organisms come first and then learn to use oxidants and reductants, or did chemosynthetic organisms adapt so that they could use photons and water instead of oxidants and reductants? Either scenario is possible, and extrametabolic reasoning must be used to choose between the alternatives.

The overwhelming source of bioenergy today is the sun, but this does not argue strongly that the sun was the original source of energy. A better approach is to ask which mode of energy utilization is simpler. We assume that the earliest life, or the chemistry that led up to life, required an energy flow, as this flux is necessary for all molecular organization. We also assume that bioenergetics preceded the emergence of enzyme catalysts because the former is required for the synthesis of the latter. Because bioenergetics existed before enzymes, chemoautotrophy is simpler than photoautotrophy, for the latter requires light-absorbing molecules that must both take up photons and then very rapidly convert the energy into chemically usable forms before it has destructive effects.

However, there seem to be a very limited number of molecular types (retinal and chlorophyll) that can absorb light energy and efficiently transduce it to chemical energy. Chemical energy transformations can take place in solutions or on surfaces and do not require photosynthetic apparatus. At the moment, all of this consideration of the first energy source is quite speculative. We are going to need some new ideas to resolve the question of what powered the earliest cells. So for now, with apologies to Robert Frost:

> Some think that life began in subterranean fire,
> And some think the start was in celestial light.
> From what I've tasted of desire
> I think that I would opt for fire.
> But light's got might
> And would work all right.

The Virtual World

T HE COMPUTER REVOLUTION has expanded and vitalized many domains in science, and has the added virtue of driving the reexamination of some important issues in the philosophy of science. To review some of these changes, consider the conventional mid-twentieth-century view, which came largely from physics. Science started with observation—the sights, sounds, colors, shapes, tones, smells, and so forth that were somehow perceived by the sense organs of the observer. From this array of sensations, one constructed objects by a procedure known as reification. One then related these constructs of the mind to others, such as mass, volume, and hardness. From these primitives of theory, one developed deeper theoretical constructs, such as molecules, electrons, light waves, and probability distributions. This set of theoretical constructs was interconnected in such a way that it led back to a set of predictions about the sensory world.

The dominant view of scientific philosophy of the latter half of this century has been that of Karl Popper: Those theories whose predictions agree with observation stand, whereas those that do not agree are discarded. For Henry Margenau, physical reality consists of the loop from sensory data through the world of metaphysically acceptable constructs back to the world of sensory data. As to some real world behind the sensory data, the *ding an sich* of Immanuel Kant, it is still unknowable.

Relating the constructs that can enter into physical theory to the raw sense data has been called the problem of epistemic connections. The word *epistemic* is related to the word *epistemology,* the branch of philosophy dealing with how knowledge is acquired. The problem of epistemic connections is still largely unresolved.

Relations between constructs and developing the predictions about observations are largely the province of mathematics, and a degree of philosophical questioning exists about how and why mathematics is such a useful tool in developing these relations. This problem was discussed by Eugene Wigner in his provocative essay "The Unreasonable Effectiveness of Mathematics in Natural Science."

With the development of high-speed calculating devices, the emphasis has shifted to new approaches that have been referred to as a "third kind of science" by Norman Metzger (*Mosaic* 16: 2, 1985). Computational science proceeds by models and simulations using theories far more complex, multivariate, and nonlinear than were possible using the kinds of mathematics I have mentioned. These models and simulations constitute a virtual world, and because of computer power one can attempt a vast number of trials in the virtual world. This generates a science of doing computational experiments in an attempt to develop large-scale structures that somehow map onto the sensory world.

Is this a new science, or is it the massive development of a previous component of the epistemologic loop of verification that now needs to be made explicit? I want to suggest that it is the latter. In retrospect, most science (which, I remind you, is a very new human endeavor) lies in the virtual world between observations (the human senses) and constructs (inventions of the human mind).

To put this in context, we go back to Athens 2,400 years ago. We may also gain a new insight into what it was that Plato and Aristotle did not like about each other. For the mathematician Plato, philosophy began with theories and proceeded to the world of virtual reality. To go from virtual reality to the world of observations meant dealing with fuzzy shadows on the wall of the cave. He was uninterested.

For Aristotle, often elbow-deep in the entrails of the animals he was dissecting, knowledge consisted of compendia of descriptions. He did not focus on the fact that the descriptions were construct-laden, so that his databases—his great biologic works— were also in the range of virtual reality. He resisted allowing theory in biology to wander far from the databases.

Neoplatonists tend to live in the world of virtual reality. The more extreme among them never challenge their models with the world of observation. Disciplines that are so oriented are always in danger of a kind of intellectual schizophrenia. Aristotelians, on the other hand, may shy away from constructs too far removed from observation (dare we call them Platonic ideals?), and are in danger of not probing the full depth of the subject. This has tended to be the approach of classical biology.

The triumph of the Galilean-Newtonian synthesis and the age that followed resulted in the world of constructs and the world of observations both entering the world of virtual reality. Thus, verification and falsification became possible for both numbers and symbols that existed in the same epistemologic domain. Because the calculations and formation of databases were pencil-and-paper operations and relatively simple, the significance of the virtual world was not stressed and sometimes not made explicit.

The existence of computational power now permits databases of great size (the human genome, large bibliographic databases such as Biosis) and simulations of great size, complexity, and sophistication (as in astrophysics, hydrodynamics, economics, and other disciplines). Thus, science does operate in new modes.

The philosophical problems in the theory of knowledge have become more difficult, if more precise. There are now two domains of epistemic connection: between theory and virtual reality, and between observation and virtual reality. I would argue that the major epistemologic questions are still unresolved.

In the world of high-speed computation, much is hidden in carrying out numeric and symbolic operations. It is important to stress that science still requires the full loop from observation to theory to observation. Failure to complete the loop results in either Platonic storytelling or Aristotelian-Baconian observation, sometimes called butterfly collecting. This is not science as we have come to know it.

The virtual world is often so much neater than the loop of scientific reality that remaining in it can be intellectually very seductive. My philosophical view is ambivalent: I am awed by the world of virtual reality as a powerful new mode of operation, and also alarmed at the power of that world to draw us away from our task of understanding the real world, whatever that may turn out to be.

Hello, Dollo!

Louis Antoine Marie Joseph Dollo was born in Lille in 1857, a year in which Darwin was at work on a book describing his view on the origin of species. Dollo, who became a leading paleontologist, joined the Royal Museum of Natural History in Brussels in 1882 and remained on the curatorial staff until shortly before his death in 1931. He contributed studies on a wide variety of taxa, ranging from dinosaurs to marsupials. His name lives on in Dollo's law, a somewhat fuzzy but profoundly important addition to Darwinism.

The contribution of Dollo is hardly a law in the formal sense of Newton's laws, and I would like to report it directly from *Evolution: Process and Product,* a standard text by Edward O. Dodson and Peter Dodson.

The evidence indicates that major evolutionary steps, once taken, are never reversed. This is known as Dollo's law. It might even be expected a priori, for major evolutionary steps are compounded of many smaller steps, each preserved by natural selection. That such a sequence, occurring by chance once, should by chance be exactly reversed would be most extraordinary. If not impossible, it is at least most improbable for whole organisms.

This sort of reminds one of that other scientist, Omar Khayyám:

The Moving Finger writes; and, having writ,
Moves on: nor all your Piety nor Wit
Shall lure it back to cancel half a Line,
Nor all your Tears wash out a Word of it.
          (*Rubáiyat*, st. 71; Fitzgerald, tr.)

In trying to discuss Dollo's law with my colleagues, I find that it has definitely receded into the past, and hardly any of the younger biologists are aware of its existence. Indeed, I would not have chosen to write about it if I had not heard the principle stated without reference to Dollo in a field remote from paleontology—the evolution of computer programs.

Returning to biologic evolution, there is a more profound consequence of the law of irreversibility than appears on the surface. We might state a corollary to Dollo's law: The more deeply a gene-encoded step occurs in evolutionary history, the more likely will a mutation at that site be lethal. Think of an example: Oxidative phosphorylation leads to the production of adenosine triphosphate (ATP), the chief energy intermediate in cells. Suppose that a mutation led to the production of XTP, and that X

could be produced with greater energy efficiency than adenosine. Since there are hundreds of energy-utilization pathways with enzymes that are strongly specific for adenosine triphosphate, the mutation would be lethal in spite of the potential ecologic advantage of energy efficiency.

Thus, the core of intermediary metabolism, the citric acid cycle, is universal because it was established so long ago in every evolutionary line. So many additional biochemical steps have been based on these intermediates that it is impossible to change them. Evolutionary change must build on what is there and cannot go back.

Consider the whales, marine mammals descended from terrestrial mammals. Whales have lungs and must periodically rise to the ocean's surface to breathe. Early in its embryonic development, a whale is a one-inch-long organism with rudimentary hind legs and the same pharyngeal arches that give rise to gill slits in fish. But in the evolution of whales, one cannot reverse the morphogenic process to have the pharyngeal arches develop into gill slits—too much has been added along the route to "mammalness" to permit the reversibility. Instead, the blowhole and other late structures develop so that the whale can spend long periods under water.

Much of this reasoning was embodied in the phrase "ontogeny recapitulates phylogeny," whose original, naive interpretation has been considerably altered in contemporary biology. Nevertheless, the great similarity between vertebrate embryos reflects the similarity of genes that operate in the morphogenesis of organisms and provide the basis on which adult forms are built.

What has all of this biology to do with computer programs? Programming also starts with some relatively straightforward rules and methods of operating that respond to computer hardware, programming experience, and user needs. As the programs come into use, modifications are made and other computing steps are continually added: Programs, like genomes, tend to grow larger and more complex.

At some stage there is no going back; the length of the programs and the idiosyncratic features create structures that cannot be reversed without in effect wiping out the programs and starting over again. In addition, as the community of users grows, so also a phylogenetic tree, an evolution of the program into its genera and species, takes root and grows.

The deep analogy between evolution and programming has gone a step further in techniques such as genetic algorithms, in which the analogues of random mutation and recombination serve as driving forces generating new programs that can be tested by fitness criteria. Here again one expects that Dollo's law will emerge.

Whether the study of evolution turns out to be of central value in computer science, I suspect we are being presented clues about the sciences of adaptive complexity. In the historical domain, survival is the game being played, and one suspects common features of all such systems. Dollo's law may be due for a revival in a much wider domain than Louis Antoine Marie Joseph ever anticipated.

Vermiforms

Among the large number of animal taxa, a substantial fraction are worms. An examination of two of the best studied groups, the aschelminths (roundworms) and the platyhelminths (flatworms), shows the profound differences that may exist among animals that look similar on first inspection yet exhibit great diversity in the way nature achieves its ends.

If I were beginning a career in biology today, I would focus on the platyhelminths, but alas, after almost fifty years as a biologist, it is a bit late to be making career choices. In any case, I offer free advice to young investigators interested in the evolutionary development of higher organisms and cognitive neurobiology: Look to the worms.

Flatworms of the genus *Planaria* are among the simplest organisms to show cephalization, the emergence of a dominant head structure. The eyes and other sense organs are controlled by a collection of neurons that can be considered the beginnings of a brain. On the great taxonomic tree of life, planaria and other flatworms appear at the branch points where two families of organisms split—the protostomia, leading to mollusks and arthropods, and the deuterostomia, leading to echinoderms and chordates, including us.

Some species of planaria have both ovaries and testes, so that when two of them mate, each deposits sperm that fertilize the other's eggs. Morphogenesis then proceeds by blastula formation in the characteristic animal mode of development. Other species of planaria reproduce asexually, with the tail coming off and growing a head, and the head growing a tail.

These organisms exhibit an extraordinary capacity for regeneration. If planaria are sliced perpendicular to the axis, the individual pieces will regenerate into complete worms. Regeneration occurs faster with pieces from the head end. If the head is sliced parallel to the axis, two complete heads form. The body plan is remarkably plastic.

Aschelminths, which include roundworms and rotifers, are also far down on the phylogenetic tree. However, I don't believe they are on the evolutionary pathway to creatures such as ourselves, since they display the phenomenon of cell constancy called eutely. As an example, in the adult rotifer *Epiphanes senta,* there are a total of 958 cell nuclei—35 in the stomach, 120 in the muscles, and so forth. The adult female roundworm, *Caenorhabditis elegans,* with a total of 959 cell nuclei, has 34 in the intestine, 111 in the muscles, and a similar fixed number in each organ system. This constancy means that within the genome there is an algorithm that precisely determines the number of cells.

Unlike planaria, nematodes such as *C. elegans* show almost no regenerative capacity and are also relatively insensitive to cancer and viral diseases. These characteristics are presumably due to the restrictive pattern of morphogenesis and tight control of cell number.

Eutely appears to be a property of organisms with no more than about a thousand cells. The phenomenon was first described in 1909. It was little studied after the 1930s until interest was revived in the mid-1960s by the selection of *C. elegans* as an organism to be investigated from a variety of perspectives. The cellular architecture of this species is probably better known than that of any other animal. There is now a complete cell lineage chart and map of the developing organism (W. B. Wood, ed., The Nematode *Caenorhabditis elegans* [New York: Cold Spring Harbor Laboratory, 1988]).

From an ecological point of view, flatworms and roundworms appear able to occupy a large range of niches. Both nematodes and planaria are found in freshwater lakes, where they live as predators or saprophytes. They appear equally able to adapt to their surroundings in spite of the differences in morphogenetic programming.

Both groups of worms are well known because of the parasitic species that cause pathological conditions in humans and domestic animals. The best known among the flatworms are tapeworms *(Cestoidea)* and flukes *(Trematoidea),* which give rise to a number of diseases. Among the roundworms *(Nematoidea),* intestinal parasites such as *Ascaris* and the causative agents of elephantiasis, *Wuncheria,* and trichinosis, *Trichinella spiralis,* are of concern in matters of public health.

As with the free-living worms, the parasites show great structural convergence in spite of having taken vastly different pathways of morphogenesis. The phenomenon of convergence speaks to the ability of evolutionary processes to produce the morphological and physiological features needed for an ecological role. That this morphological versatility exists in animals under such precise developmental control as the roundworms, as well as in the highly plastic flatworms, is particularly surprising. It seems clear that it represents an evolutionary feature that emerged early in the history of multicellular animals.

Whose Standards?

IT WAS both surprising and pleasing to come across the article "SI Units: Wrong for the Right Reasons" by Allen B. Weisse (*Hospital Practice,* December 15, 1992). On reading the article, I realized that Dr. Weisse and I are not of one mind on standard international units, and an answer is required.

His essay on SI units takes off on an introductory theme: "Do you know *anyone* who has ever studied Esperanto?" The message, of course, is that some activities, while theoretically reasonable, are just impractical. However, I need to reply to that question. For two years I was faculty adviser to the Esperanto Club of George Mason University, whose chief activity was to offer a course in Esperanto. I am now possessor of the book *Esperanto,* by David Richardson, which is the source of most of my material on the language.

I do not speak this international tongue and was able to attend only one class, so my knowledge is shallow. The students who asked me to serve as faculty adviser were of the Baha'i faith, a group that learns Esperanto for worldwide communication among the faithful. I replied that English is the international language. They argued for a less politically embedded international tongue, and I agreed to be the adviser as it required very little of me. The classes now seem to have been abandoned, and I suppose my fellow columnist is right: Esperanto is not the wave of the future.

While yielding on Esperanto, my passion on standard international units will not be assuaged. A lifetime of research and teaching does not allow my usual moderation when it comes to dealing with units. I even remember when my conversion occurred. Because of an interest in molecular motors in bacterial flagella, I set out to calculate the smallest electromagnetic motor that could be built. It was a harrowing calculation involving expressions from mechanics, electricity and magnetism, and hydrodynamics. The final expression had about a dozen parameters from those fields. The calculation was driving me bonkers, because almost every parameter was found in the literature in its own set of units and it was necessary to rationalize all of them.

The moment of truth came when I realized that if I simply put all the quantities in SI units, I did not have to worry: All problems relating to units would take care of themselves. Not exactly a burning bush or a fall on the road to Damascus, but it was a profound intellectual experience that has indeed been of lasting influence. I was in the final manuscript stage of *Foundations of Bioenergetics,* and much to my proofreader's disgust, I went back and recast the entire book using SI units. (This was no trivial matter, as my proofreader was also my spouse.) Now that I have been "saved," I have a missionary zeal for this international scientific convention.

The story of standard international units goes back to Paris in 1795. The French National Assembly adopted the meter and the kilogram based on physical standards created and maintained by the Paris Academy of Sciences. On May 20, 1875, seventeen nations, including the United States, signed the Convention of the Meter, which provided for an International Bureau of Weights and Measures.

In October 1960 at the Eleventh General Conference of Weights and Measures, the International System of Units was adopted, to be designated as SI in all languages (I assume this includes Esperanto). The system was based on the meter, kilogram, second, ampere, kelvin, and candela. Subsequent conferences have refined the standards, adding the mole as a primary unit, and appending the radian and steradian as supplementary units.

There are a number of derived units, most of which are named after distinguished scientists. The one that caused us the most trouble for the thermodynamics book was the unit of pressure, the pascal. It is of course equal to 1 newton per square meter. The former is the force exerted by a kilogram mass at the earth's surface (where the acceleration of gravity is 9.8 meters per second squared). Since force is mass times acceleration, a newton is $9.8 \text{ kg-m}/\text{s}^2$. Thus, an atmosphere is 101.325 kPa. It is a bit confusing at first, but all very rational.

I must confess that a blood pressure of 16,110.7 Pa sounds very strange, but in the days of digital readout, are we to be forever constrained by the fact that sphygmomanometers once used columns of mercury: hence, in this case a reading of 120/80 millimeters of mercury? Perhaps we are moving the discussion to the question of the art of medicine versus the science of medicine, an arena I am clearly unqualified to enter. I lost an argument over this in 1952, and I've been reluctant to argue the point since.

Back to my beloved SI units: A series of prefixes now covers a range of 10^{36}. Thus, for example, 10^{-18} meter is an *atto*meter (a very small distance indeed) and 10^{18} meters are an *exa*meter (about 100 light-years). For many years certain wags in my laboratory would point out that a 10^{-9} goat is a nanogoat. All of this is to introduce an invaluable resource, *The International System of Units* (E. A. Mechtly, Stipes Publishing, 1977). New versions may be available, but this edition is especially easy to use.

One of the greatest values of using SI units is in teaching physics and chemistry, especially physical chemistry. In the good old days, we would spend countless hours doing problems that amounted to nothing more than converting from one set of units to another. This was almost pointless, since it teaches nothing about the subject itself but deals with convention adopted in different countries at different times for different reasons. This time can now be spent on content. With respect to medicine, I would argue the same standards, at least for biochemistry. Thus, joules should replace calories, and I hope this will carry over to metabolism and physiology.

Before ending this sermon on SI, I should report a conflict of interest. Many years ago I spent two years working at what was then known as the National Bureau of Standards. I also want to make peace with traditionalists who think I may be doing violence to the past. I reconcile these problems by noting that one teaspoon equals exactly $4.92892159375 \times 10^{-6}$ m^3 by definition. And finally I wish to note that a cubic meter is a stere: a fact known to crossword puzzle aficionados and almost no one else.

IV

The Ecosystem

Atoms and Ecosystems

Ｔｉｔｕｓ Ｌｕｃｒｅｔｉｕｓ Carus was one of the last members of an intellectual tradition that spanned four hundred years. In an ancient genre, he wrote a scientific treatise that was also a poem "with much art," as noted by Cicero. This combined science and poetry was clearly the antithesis of C. P Snow's two-culture dichotomy and represented an attempt to unify mankind's thoughts about the world. *De rerum natura,* written in Latin and rendered in English as *On the Nature of Things*, dates from 55 B.C., which was also the year of Lucretius's death.

The work of the Roman savant is extraordinary for its time and perhaps for ours, for it begins with a plea to give up the superstition that guides men and to seek for knowledge—science—so that by understanding the causes of phenomena we will not, in terror, resort to the irrational. His science derives its core, through Epicurus, from the great classical atomists Leucip-

pus and Democritus. And so the poem begins by announcing
two grand unifying principles:

- Nothing can ever be created by divine power out of nothing.
- Nature resolves everything into its component atoms and
  never reduces anything to nothing.

Together these constitute a conservation law and a terse statement of the atomic hypothesis. In his introduction, Lucretius
called atoms primary particles "from which nature creates all
things." While modern historians of science have long marveled
at the perspicacity of the early atomic hypothesis, I have not seen
a discussion of the remarkable fact that Lucretius used these ideas
to infer biologic recycling. To get a sense of both the poetry and
the understanding of the recycling process, we turn to the translation by R. E. Latham (Penguin Books, 1951):

> Lastly, showers perish when father ether has flung them down
> into the lap of mother earth. But the crops spring up fresh and
> gay; the branches on the trees burst into leaf; the trees them-
> selves grow and are weighted down with fruit. Hence in turn
> man and brute draw nourishment. Hence we see flourishing
> cities blest with children and every leafy thicket loud with
> new broods of songsters. Hence in lush pastures cattle wea-
> ried by their bulk fling down their bodies, and the white milky
> juice oozes from their swollen udders. Hence a new genera-
> tion frolic friskily on wobbly legs through the fresh grass,
> their young minds tipsy with undiluted milk. Visible objects
> therefore do not perish utterly, since nature repairs one thing

from another and allows nothing to be born without the aid of another's death.

Thus, the notion of cycling is introduced almost casually, and the poet goes on to another phase of his argument, that nature consists not only of atoms, but of empty space within which the atoms may move.

At one point, Lucretius reminds his readers of his reason for using poetry:

Our doctrine often seems unpalatable to those who have not sampled it, and the multitude shrink from it. That is why I have tried to administer it to you in the dulcet strains of poesy, coated with the sweet honey of the Muses. My object has been to engage your mind with my verses while you gain insight into the nature of the universe and the pattern of its architecture.

Herein lies a lesson for those who wish to convey science to a broader public.

Later, Lucretius restates the theory of ecological cycling:

Be sure that matter does not stick together in a solid mass. For we see that everything grows less and seems to melt away with the lapse of time and withdraw its old age from our eyes. And yet we see no diminution in the sum of things. This is because the bodies that are shed by one thing lessen it by their departure but enlarge another by their coming; here they bring decay, there full bloom, but they do not linger there. So the sum of things is perpetually renewed. Mortals live by mutual interchange. One race increases by another's decrease. The generations of living things pass in swift succession and like runners hand on the torch of life.

He returns to the argument on other occasions, but always the decay of all things and the indestructibility of atoms keep forcing Lucretius to the idea of cycles. Once again, ideas believed to be newly formulated are deeply embedded in our cultural history. A book written more than two thousand years ago is not where one usually goes for references on ecology, yet there it is in clear poetry: the basic fact of environmental dynamics.

While rereading the defense of Epicurus by Lucretius, I am reminded of the bum rap that history has given the early Greek philosopher. I remember when I was twelve years old and attended synagogue with my grandparents. The culture was displaced Eastern European, and I understood only a little of the Yiddish. Yet I recall a word of extreme moral censure: *appikoros*. It referred to one of loose life, one who was lascivious. It was almost always pronounced scornfully from within the walls of the house of prayer.

Only years later did I realize that this was the same Epicurus immortalized in the English word *epicure*. How did knowledge of this Athenian philosopher, imperfect though it was, find its way to the Lithuanian, Polish, and Russian villages where my grandparents' peers had lived? I suppose it must have begun in that interaction between Jews and Hellenists in Alexandria at the time of Lucretius and then found its way along the route from North Africa to Western Europe to Eastern Europe. This sixteen-hundred-year journey preserved knowledge of Epicurus's love of life but not his temperance, serenity, and love of culture or his role as an intermediate between Democritus and Lucretius. Then again, there wasn't much discussion of atomism at the synagogue.

This misunderstanding of Epicurus has been pretty universal in Western culture. Whether we can trace recycling from Lucretius back to his forerunner in Greece is a worthy matter for scholars to consider. In any case, we can agree that recycling is a corollary of the permanence of atoms.

The Technosphere

THE GREAT DEVELOPMENT of geologic knowledge in the nineteenth century gave rise to the concept that geochemistry could be formulated in terms of the properties of and exchanges between the geospheres: lithosphere (earth), atmosphere (air), and hydrosphere (water). In the beginning of the twentieth century, Vladimir Ivanovich Vernandski turned his attention to the role of living organisms in interaction with the geospheres. Thus began the science of biogeochemistry. This viewpoint led to the notion of the biosphere, the grand total of living and dead material on the planet. The biosphere has been a useful concept and is now part of mainstream geochemical thought. Although its mass is small, it has a great catalytic effect on the three other geospheres and thus assumes significance disproportionate to its size.

The study of earth-system sciences over the past twenty years shows that the four-geosphere concept is incomplete if we are to study such issues as the greenhouse effect, the composition of the atmosphere, the movement of minerals and other elements from land to ocean, and the regulation of the upper-atmosphere ozone concentration. For in these problems we encounter anthropogenic perturbations of major proportions. The changes cannot be accounted for within the biosphere as envisioned by Vernandski but must be attributed to another geosphere—the *technosphere,* the total of all human engineering facilities and activities. Thus, for purposes of environmental policy, the earth must now be envisioned as a group of five interacting geospheres. Nuclear power plants, agriculture, and massive burning of fossil fuels are so different from normal biologic activity that they require a new sphere of their own.

In principle, the origins of the technosphere go back to the beginnings of toolmaking and thus could probably be linked with the appearance of *Homo habilis* two million years ago. However, for as long as the genus *Homo* was a hunter-gatherer, its activities could be encompassed within the biosphere. Beavers build dams, termites make mounds, and corals form reefs. The biosphere is not without its engineering components.

Something of a different order of magnitude was introduced with the agricultural revolution ten thousand to fifteen thousand years ago: There began a conversion of plains and grasslands from complex ecosystems to virtual monocultures. The biosphere was being altered in major ways because of anthropogenic activity. This alteration of the biosphere had significant effects on the other spheres because it is, as noted, such a catalytic component. The evolution of the planet was no longer predictable from the laws of physics, chemistry, and biology alone. Beginning with the rise of agriculture, activities of human societies had to be taken into consideration.

All of this happened so gradually that the technosphere, including fields, roads, and bridges, seemed as part of the natural landscape. The pastoral scene is not nature at all but entirely the result of human technological activity. One is reminded of the anecdote about the nineteenth-century romantic who commented: "Wasn't it wonderful of the Lord to have rivers flow through all the major cities of Europe?"

With the Industrial Revolution, there was a huge increase of technospheric activity. The mining of fossil coal and the drilling for fossil oil released the energy and mass of these biospheric components into the atmosphere and hydrosphere.

Because the magnitude of this fossil-fuel conversion grows every year, the technosphere is having an effect on atmospheric composition and climate. Clearing land; building dams, canals, and dikes; and constructing roads and tunnels affects mineral flow from land to rivers to oceans. This in turn affects biospheric components.

Just as the origin of life led to a biosphere that profoundly affects many aspects of the planet, so the origin of the genus *Homo* gave rise to a technosphere, a geosphere of profound significance. Although *Homo sapiens* is clearly a biologic organism, the technosphere is not part of the biosphere. For the evolution of mankind resulted in mind as a major feature, and mind eventually gave rise to engineering activity that greatly exceeded the biospheric potential of humans—hence, the technosphere.

At present, we face the consequences of habitat alteration, global climate change, and loss of biodiversity. These are clearly the result of the technosphere's interaction with the other geologic spheres. Some people view these new features as terrors of human activity. One can generate a little calmer reaction by simply studying the consequences of a growing technosphere. It is now clear that the new sphere has achieved major importance in the geochemistry and geophysics of the planet and should be admitted to the geologist's collection of significant components.

During discussion of the term *technosphere* with colleagues, some suggested two other terms that they thought more appropriate. There was *noosphere,* a term that stresses the fact that we are dealing with the results of the uniquely human mind. However true this may be, *mind* seems a somewhat ethereal construct for belching smokestacks and huge dams. *Noosphere* had best be reserved for more spiritual ventures.

The second suggestion was *econosphere,* because the rise of engineering activity accompanied the rise of economics as a human activity. Econosphere, which suggests the roots of human effects on the other geospheres in the economic activities of *Homo sapiens,* falls between *noosphere* and *technosphere* in stressing the mental roots of the physical changes occurring on the planet. In that sense, it is a telling term.

Nonetheless, I stick with *technosphere,* for geologists are not a likely constituency to argue the mind-body problem, and we can study planetary effects within the more materialistic technosphere. There is a real social component in all of this, however, and it is well to remember that the geochemical future of our planet does depend on what goes on in our minds. It also depends on the number of minds on the planet, each of which is associated with a resource-consuming body.

An Ecological Test Tube

Eearly in the morning of September 26, 1993, eight individuals emerged from a 3.15-acre closed ecosystem into the open air of Oracle, Arizona. Their emergence after two years inside was an event that I believe will be remembered in the history of ecosystem ecology.

Associated with the first two years of Biosphere II are a number of interweaving stories. The one I choose to tell is the scandal of ongoing harsh criticism of the project by the media and segments of the scientific community who fail to understand even minimally the conceptual basis of what was being attempted and accomplished.

First, some background about the scientific concept of closure. Consider a stoppered test tube. The interior is closed to the flow of matter while a reaction is taking place, but energy in the form of heat may be taken up or given off. We use a test tube to isolate the contents from the surrounding materials so we can concentrate on what is happening to the materials inside. Similarly, a bacteriologist using a tube of sterile nutrient broth stoppered by a cotton plug deals with a system open to the flow of heat and atmospheric gases but closed to the entry of bacteria, except those placed there on a sterilized metal loop. The story of science is replete with cases of closing off systems in order to study them under controlled experimental conditions.

Some sciences, by the very nature of what is being studied, do not permit closure and are thus observational. Astrophysics is perhaps the clearest case. There is no suitable enclosure to put around a galaxy so we can perform experiments on it. Chemistry, on the other hand, is a science of closures, and many types of complex glassware have been developed to allow manipulation of the contents of closed systems.

Traditional ecology has been primarily observational, taking place in fields and forests, oceans and reefs, rivers and streams. Some theoretically oriented ecologists have sought the partially isolated systems of lakes and islands as a route to partial closure that has permitted some useful experiments.

Biosphere II sought to design, construct, and operate a closed system to serve as a laboratory for ecosystem research, with closure defined as not allowing the flow of matter but permitting the flow of energy and information. This condition approximates that of our planet, which receives energy from the sun and radiates it to outer space while being almost closed to the flow of matter. Occasional meteoroids arrive and some hydrogen leaks out, but this mass flow is small, even though it may have informational importance. In any case, the general condition of the earth has been summed up in the dictum "energy flows and matter cycles."

The first two years of Biosphere II were an engineering and biologic test of a closed large-scale complex ecosystem operated by a crew of humans who had to survive on the internal material resources; therefore, they recycled water and waste, and they maintained an intensive agricultural system.

Almost none of the journalists who criticized the project understood the facts just outlined. They avoided this introductory education and wrote without knowledge, their comments ranging from the acerbic barbs of the *Village Voice* to the bland know-nothing editorials of *The New York Times*.

Oddly, many scientists relied on this journalistic sniping as the sole basis on which to develop their own opinions of the validity and worth of the project. When I first attended a conference at Oracle two years ago, some of my scientific colleagues asked, "Are you going out there to associate with those———?" I replied, "I am an experimental scientist. I will not judge a project until I get information for myself." My answer still stands. The firsthand information I got convinced me of the validity of the data being obtained and the integrity of those obtaining it.

I ask the many scientists who have expressed negative remarks based on snippets of coverage by the mass media: Would you like your own work to be evaluated by reporters who have not done their homework and are looking for the excitement of a scandal?

Books can and should be written about the relationship between this project and the media, whose coverage was both unimaginative and silly. For example: Fifteen days after closure, Jane Poynter suffered a finger injury that required treatment by a surgeon. She exited and reentered the Biosphere by a double-interlock door and was outside for about five hours. She returned in full view of the media with a duffel bag containing computer and scientific supplies and maps. At the very most, this event could be considered to have reduced the closure period from 104 to 102 weeks. Even if the bag had been full of food, this one event would have altered mass balance by at most a minute percentage. The incident did not compromise any part of the study, yet we continually hear about it. In the process, focusing on—and distorting—the irrelevant, the media have missed an important story, the beginnings of a new kind of ecosystem study that may be of major significance in the unfolding of our understanding of biology.

A small number of other irrelevant tales also received constant repetition. Nothing new was added. The harassment went on, overlooking a wealth of interesting technical features. I suppose the media were doing their thing, but what about the scientists who accept this type of information as the basis for their judgments? That must be peer review at its very worst.

Ficus

SOMETIMES MINOR EVENTS unfold in strange ways, thus leading to reflection. For reasons I cannot fathom, I have always been fascinated by Buddha's decision to sit under the bo tree until he received enlightenment. That particular tree was a fig, a member of the species that by the time of Linnaeus had acquired the scientific appellation *Ficus religiosa*.

Recently, I was in Lahaina, Maui, a location that evokes thoughts of Buddha because just outside of town at a shrine and temple is a large bronze statue of the Enlightened One. From time to time I have stood in front of that monument thinking about enlightenment.

A friend of mine has, as a volunteer, created an extraordinary garden near Lahaina town center. In the midst of the greenery is an ornamental fig tree; I believe it is a *Ficus benjamina*. I half jokingly suggested that we should put a sign under the tree in both English and Japanese with the message "Reserved for Buddha." I then began to wonder about the relation of this species of fig to *Ficus religiosa*. A short distance away was the public library, where an encyclopedia article surprised me with the information that there are more than eight hundred species of figs, including the banyan, a magnificent specimen of which was shading an entire square city block just a few yards away.

All this talk of figs led me down Prison Street to Dan's Green House to see what species were being sold. I was shown first some bonsai trees growing out of porous lava and was immediately charmed by the mistletoe fig, *Ficus deltoidea* (also known as *Ficus diversifolia*). I left town with my very own miniature tree. It was becoming increasingly clear that when the bonsai found its appropriate place in our residence, a small statue of Buddha should be seated under its branches. I returned home with two missions: to find a suitable representation of Siddhartha and to learn a little more about the genus *Ficus*.

The first mission proved easier than the second, and a small cast bronze Buddha with a patina finish now sits under the tree. That search was not without its own story. I was in an antique shop in northern Virginia and saw a large Buddha. I asked the clerk, who was Asian, if she had a smaller one. In halting English she said it was the only one she had. "I was once a Buddhist and had several hundred statues," she went on. "Then I converted and became a Methodist and sold all the Buddhas." Such is the way of the world.

My search for enlightenment about the genus *Ficus* had its difficulties. My encyclopedia at home asserted that there are more than two thousand species of figs, indicating at least some disagreement among experts. I did find that the common commercial fig is *Ficus carica*. It has been a food source since antiquity and is of major agricultural significance in a number of Mediterranean countries and the United States. The relationship between the fig wasp *(Blastophagus psenes),* the caprifig *(Ficus carica sylvestris),* and the commercial fig is a complicated story that I shall refrain from detailing, save to point out that it has lessons to teach us about coevolution.

Figs are of cultural and religious importance in the West as well as the Orient. In the Abrahamic religions, fig leaves were made into the first clothes by Adam and Eve when they became aware of their nakedness in the Garden of Eden (Genesis 3:7). Artists of the Renaissance frequently used large fig leaves to avoid depicting genitalia. Figs were sacred to the Greek god Bacchus and were used in religious rites. And the story of Romulus and Remus has it that a fig tree shaded the entrance to the wolf cave where they lived.

The quest for more detailed information about the biology of *Ficus* led me to an extraordinary volume, *A Monographic Study of the Genus Ficus from the Point of View of Economic Botany,* by Nagaharu Sata (Taihoku Imperial University, 1944). Part I is about the economic botany of Formosan species; part II deals with systematic studies of Philippine species; and part III is a hidden scholarly gem, "Historical Studies of the Genus *Ficus*."

Part I contains a description of *Ficus diversifolia*: "The plant is very fruitful even when it is planted in a little pot." Of *Ficus reli-*

giosa, Sata concludes, "The story, much told as it gave a shade to Buddhah who was sitting under the tree, also favors the planting of this plant in the orient."

In part III, a biological scholar's paradise, it is evident that the author has taken his subject to heart. He has read every work and examined every species available to him, and he has thought about the significance of this knowledge in terms of systematics, the theoretical bases of systematics, the origin of cultivated plants, the relation of *Ficus* to the family Moraceae, plant morphology, and geographic distribution and evolution. He outlines the history of *Ficus* classification from the seven species recognized by Linnaeus to the 1,580 species recognized by the Index Kewensis in 1930. Sata estimates that there may be as many as three thousand species once all have been discovered.

Owing to Sata's enthusiasm for his subject, I now know more about the genus *Ficus* than I ever dreamed of learning. Be assured that I have resisted the temptation to share with you more than a tiny fraction of the vast botanical and anecdotal lore I now possess. (Would you like to know about the ornamental rubber plant *Ficus elastica* or the sycamore *Ficus sycomorus*?) This knowledge will be of no value to me in the marketplace, but I look at my Buddha and think that enlightenment has many forms. However, the next time I hear someone use the phrase "It's not worth a fig," I may be tempted to take that person aside and have a conversation.

Apocalypse

E very so often, a group of astrophysicists will release more detailed information to remind us how much or how little time our planet has left before a celestial apocalypse reduces us to ashes. The latest estimates indicate that the Earth has been around for about 4.5 billion years and will be totally uninhabitable in about 1.5 billion years. Since life has been around about 4 billion years, it has run 73 percent of its course. From a cosmic perspective, it's later than you think.

Whenever this subject comes up, I am reminded of a fellow graduate student, Alex Mauro. Shortly after my new bride and I had settled into our apartment, we invited Alex over to dinner. My friend seemed somewhat distraught, so I finally asked what the problem was. He replied that he had been thinking about the Second Law of Thermodynamics, and the thought of the universe decaying to entropic doom was profoundly discouraging. I have made it a policy in life to concentrate on things I could do something about, and the Second Law seemed clearly outside of that domain. My efforts to cheer up our guest were not very successful. Lucille's cooking was a more effective antidepressant, and Alex left in a much better mood.

Much has been learned since those graduate school days, and astrophysicists have been able to attain a fairly detailed understanding of the life and death of stars. Therefore, the predictions about the future of our planet are on a very sound scientific footing. Long before the Second Law will do us in, Earth will be incinerated when the sun becomes a red giant.

That certainly brings the end much closer, but since it is still 50 million human generations away, it hardly affects me personally. As it happens, other apocalypses may arrive much sooner. For example, it is now fairly well accepted that the enormous changes in speciation at the Cretaceous-Tertiary boundary 65 million years ago were caused by a very large meteoroid striking the Earth in the Yucatan peninsula. We do not know when or if another large celestial body will strike the Earth, nor can we predict the consequences without knowing its size and where it will hit. But the possibility exists.

There are, of course, other apocalypses that may even be closer. Certainly, nations possess enough nuclear weapons to wipe out human life on Earth, and given our fallibility, we have no guarantee that those weapons will not be unleashed. Controlling the human capacity for self-destruction is a task that we must keep working on for the foreseeable future

Now, I don't want you to think that your neighborhood optimist has suddenly soured on life, but I do want to muse about the impact of apocalypses in forming philosophies. We must start with the fact that all systems arise, flourish for a time, and perish; that is the existential nature of everything we know. Therefore, to focus on the indefinite future in the search for meaning is contrary to what we understand. We must turn to the flourishing as our focus for life philosophies.

That flourishing gives us plenty to think about. For in everyday existence, there are, as Buddha stressed, problems and pain and suffering. We need not focus on some distant apocalypse but can turn our attention to the four horsemen that are running about today: sword, famine, pestilence, and wild beasts. Until we have conquered war, starvation, epidemics, and man's inhumanity to man, it seems gratuitous to invest emotion in our individual or cosmic transience.

War is probably the most inane human activity. Yet war has been part of our heritage ever since we appeared as a species. Indeed, our emergence may have been achieved at the price of genocide: It has been suggested that our ancestors killed off competing species of related hominids. Given the present state of technological sophistication, one would think that large-scale societal killing would stop, but it doesn't. That must be the apocalypse within.

Starvation is a horror we should be capable of dealing with, yet that too has a Sisyphean aspect. With each new advance in agronomy and food production, the growth of population, war, and our inability to distribute food leaves malnutrition and starvation as major problems. The population explosion leaves us asking, "How crowded a world do we wish to live in?" Again, no consensus seems to be forthcoming.

Pestilence is a horror that we thought we had almost banished, but it appears to be more deeply embedded within the nature of organismic biology than we had anticipated. Antibiotic resistance is constantly outstripping the discovery or synthesis of new antibiotics. The appearance of new viruses and other infectious agents leads to diseases unknown only a few years ago. The HIV epidemic certainly restrains us from feeling complacent about pestilence.

The problem of wild beasts has changed in character since the Book of Revelation, which concerned itself with the fear of wild beasts running rampant. We now concern ourselves with the fear of wild beasts driven to extinction. For hunter-gatherers, wild beasts were a constant danger: a grizzly bear along a trail in the Rockies, a lion near the Tsavo River in Kenya, or a crocodile on the shore of Lake Turkana. These fears were very real and much different from the abstract angst about anthropogenic changes resulting in destroyed habitats and extinction of animal species. Species disappearance of both animals and plants stands out to some as a modern apocalypse.

Well, if Alex were alive today, I would say to him: "Look, friend, apocalypses happen all the time. That's the kind of universe we live in. We cannot base our philosophy on a concern with future catastrophes; the present exigencies are more than enough to contend with. Plan for the future, of course, but only for the foreseeable future. As individuals, we will occupy at most a tiny fraction (1.5×10^{-8}) of the life of our planet. Let's make the best of that fraction. Concern with the forever is hubris."

Krill

As one interested in the great whales, I have thought a lot about krill, but I met my first *Euphausia* only a few months ago on the bridge of the *S.S. Hanseatic* as the ship was pulling up anchor off one of the South Shetland Islands near the Antarctic Circle. This was my first trip through these waters, and I was once again marveling at this ecosystem that manages such productivity at very low temperatures. The ship's engineer appeared on the bridge with a bottle of seawater from the engine's cooling system. The water was teeming with very small crustaceans, probably a hundred in the pint-sized jar. These were the famous krill.

We had for the past few days been seeing the plentiful large-scale fauna of these seas and shores: penguins, skuas, albatrosses, petrels, seals, and an occasional whale. These large forms subsist because of a remarkable food web adapted to this seemingly hostile environment. The krill form a key component of this network.

As in all photosynthetically driven ecosystems, the primary producers are green plants, which take up water and carbon dioxide to form biomass. In temperate lands, the producers are trees, grasses, and shrubs. In the Antarctic, even in summer, I see almost no green, mainly the white of snow and ice and the curious deep blue of the cracks in the glaciers. Where then are the primary producers? They are too small to see individually. They are the phytoplankton growing in the sea and the ice-associated algae. All food in the Antarctic ecosystem must ultimately derive from these organisms working within the limitation of very short winter days, which must be compensated by growth in very long summer hours of sunlight.

The reason sizable amounts of algae materials do not pile up is that they are rapidly eaten by zooplankton or krill. There is thus a very rapid turnover of primary producers, and these algae must grow rapidly to keep supplying food. During the summer, with almost constant sunlight and few competitors for primary nutrients, the diatoms and other algae do grow rapidly, thus providing fixed energy for the entire ecosystem. Generation times are on the order of days in the icy cold of Antarctic offshore waters.

The primary production ends up in algae, grazers that eat the algae, and decomposers (bacteria that pick up nutrients released into the water). The krill are omnivores with a feeding and digestive apparatus that allows them to eat the algae, the diatoms, the zooplankton, and the bacteria. A large fraction of the total energy of the system thus flows through the krill, mostly of the species *Euphausia superba*.

The krill, which are central to this ecosystem, are then eaten by baleen whales, squid, seals, birds, and pelagic fish. The squid are eaten by sperm whales, seals, and birds, and the pelagic fish are eaten by birds. The feature of a highly seasonal ecosystem of this type is the necessity to store up a lot of energy in summer. It has been reported that adult whales and penguins may lose half of their body weight in winter. A large baleen whale feeding in summer may eat up to three tons of krill per day. These ecosystem flows may be described as all of the small organisms being eaten by krill, and krill being eaten by all the large species. It is rare for a single species to play such a crucial role.

Traveling through the South Shetland Islands provides one an opportunity to see the richness of biota. The most photographed examples are the penguin rookeries, where thousands of birds may be seen nesting on curious nests built up of small stones, the most available material for the purpose. The ubiquity of penguins characterizes not only the travel posters but many of the locations we had a chance to visit.

Another example of the richness of life south of 60°S latitude is the large number of small icebergs, each with a single seal sunning itself on its floating temporary home. There are five species of seals in Antarctic waters, with rather different diets of krill, squid, fish, and other prey species.

Leopard seals, which are among the most aggressive predators, have a very different diet over the year. In September, krill constitute 60 percent of their diet. By January, they mostly eat squid, penguins, and other seals. In February, penguins account for 30 percent of their diet. During the Antarctic winter, they largely depend on the marine food source of fish, squid, and krill. Leopard seals also prey on other seals, mainly crabeaters. The Antarctic food web is highly interconnected.

Whales are of two general groups: toothed whales (like Moby Dick) and baleen whales, which filter seawater through the baleen structure and extract krill for food. Herman Melville referred to the baleen apparatus as Venetian blinds. While the filter feeders are clearly dependent on the krill, so are the toothed whales by a somewhat longer food chain. Krill constitute one of the main foods of the Antarctic squid, and these squid are a major food source of the toothed whales.

There has been talk lately of harvesting krill and converting them directly into human food. Some ecologists worry that taking such a central element from the food web would be destabilizing. The Antarctic ecosystem seems so highly specialized that I, too, take a conservative approach to a key node in the network. Are we good enough ecosystem modelers to predict the effects of major perturbations? I don't think so, but in the meantime, the system is an excellent one for modeling studies because the total number of species is quite small, there are major barriers to migration, and the seasonal effects are so large. Before that modeling is thoroughly understood, I am prepared to forgo krill stew.

Arks and
Genetic Bottlenecks

A FRIEND SENT ME an article that addressed the search for an environmental ethic in the biblical background of the Judeo-Christian tradition. Although the article did not mention it, I began thinking about the major ecological disaster treated in the Bible, the Noachian flood. Most of the planet's terrestrial species were reduced to two members, one male and one female. The rest of the organisms were drowned by the rising waters that covered the earth. This flood, which drove all species to the brink of extinction, resulted from the evil actions of the people of the planet. Moral number one: If we, the inhabitants of the planet, do not behave ourselves, we jeopardize the welfare of all species.

But, alas, Noah must have misunderstood the Lord, for two is most likely an insufficient number of individuals to save a species from extinction after the waters recede. The difficulty has been given the name "the genetic bottleneck." It states that if a species is reduced to a small number of individuals, the amount of genetic variability is so reduced that it is unlikely that the species possesses enough phenotypic variability to survive environmental changes.

Consider that a species is a collection of interbreeding organisms. No two individuals, with the possible exception of identical twins, have the same assortment of genes. Indeed, every individual has paired chromosomes, with two copies of every gene. Often these paired genes are not identical. Nonidentical genes for the same trait occurring at the same locus on a chromosome are designated alleles. The number of allelic forms of a gene may be as few as one or as many as ten or more. All individuals are different as a consequence of the allele(s) they possess at each locus on their chromosomes. A species is therefore both a collection of genetically varied, interbreeding individuals and a common pool of genes that are available for future generations to inherit.

The gene pool, with all of its variability, is considered to be essential to the survival of a species, because one of the trials that all organisms face is a varying environment. This is due to changes of climate as well as to competitor or predator species invading or arising in the habitat. Geological factors due to tectonic shifts, rifts, or other changes may also alter a species' surroundings.

The fitness of a given individual depends on its genome—that is, on which genes from the allelic set it contains at each locus.

The more variable the gene pool, the more likely it is that some members of the species will survive any environmental insult, regardless of how severe it is. Conversely, an extremely nonvariable gene pool seems unlikely to result in enough individuals to make it through a crisis.

As a result of these features, most population ecologists feel that when the number of members of a species drops below some critical level, survival in the wild is not possible; intervention, captive breeding, and protected habitats are considered absolutely necessary for survival. Cases in point are the Hawaiian nene and the California condor.

Let us consider the habitat faced by the animals as they step off the ark on Mount Ararat. The receding waters leave little in the way of vegetation for the herbivores. Even if seeds survived the soaking, it will take a long time for plants to grow to the point of providing sufficient food. There are going to be some very hungry herbivores around.

For the rest—the carnivores and the grazers they eat—the situation is going to be even worse. After the hungry lions have eyed the Thompson's gazelles, there are going to be extinct Thompson's gazelles and still hungry lions, who may then turn their attention to the two gnus. Eventually, this could lead to a major species extinction, as a hungry lion has to eat a lot of starving herbivores to keep going. Why didn't the lions evolve into herbivores? With such a small gene pool, they couldn't. It is thus extraordinarily unlikely that the lions could have survived after leaving their temporary home. No, I'm afraid the scriptures don't offer much help on the biodiversity problem.

This may be part of our difficulty. We generally operate within an ethic and morality that took shape in a world of relatively low human population, when infant mortality was a major problem and "Be fruitful and multiply" sounded like very good advice. Nowhere do we find hints of biblical overpopulation, although famine was clearly an ever-present threat. Biblical morality is very much about the relationship between individuals or relatively limited groups of individuals. If the Bible says little or nothing about current large-scale problems, it's because they didn't exist in the times of Noah, Moses, or Jesus.

Thus, we now have a world of habitat destruction driven by population pressure, and we are yet urged to be fruitful and multiply, prompted by some religious leaders to reject contraception. We are caught in a time warp, living in one kind of world with an ethical system from an entirely different kind of world. I don't think we can just go back to the good book(s). We're going to have to figure it out for ourselves.

V

Criticism

A Long Line of Reasoning

It IS EARLY AFTERNOON, and I am sitting quietly at the kitchen table sipping tea and reassessing the events of the past two days. The happenings were neither unique nor all that exciting, yet they deserve careful thought for what they may have to say about life in the United States. This piece may be regarded as catharsis to calm the troubled soul after I took part in a mass exercise in the Tao of patience presided over by the State of Connecticut.

A few days ago, in a surge of madness and an attempt to revive the sagging economy, we acquired a 1983 Chevy station wagon, and the task fell to me to register the vehicle. Accompanied by my grandson and fully equipped with the bill of sale, title, current registration, insurance certification, and a registration application, I drove to the Hamden office of the Connecticut Department of Motor Vehicles. The first thing to be noticed was the number of people and cars converging on this place. My grandson thought the scene looked like Shea Stadium, but considering the way the New York Mets have performed, the crowd may have been even larger.

Inside, the first station was the information desk. One does not, of course, immediately go to "information": One goes to the end of a serpentine line of seventy-nine persons wishing to approach the first desk. The purpose of visiting this desk is to have one's documents examined and, if they pass the test, to receive a serial number for moving to one of the windows for further processing. I hesitate to think what happens if one does not pass this first qualifying test, but fortunately I made it and received a ticket with the digits 585. High in the air was a display panel lit up with the digits 530. Clearly, a period of enforced patience was being presented. We strolled, we sat, we ventured across the street for coffee and doughnuts. We regretted not having brought a book or two.

Slowly and inexorably, the numbers rolled by at a rate of slightly less than one per minute, so about an hour later, we arrived at the exalted position at the head of the line and approached the second station, the designated registrar's cage marked by a lighted bulb. On negotiating with the representative of Connecticut, I found I was being issued temporary license plates that permitted me to drive the car for twenty days and within that period to appear for a safety inspection required of superannuated vehicles. That seemed reasonable enough, and led to the following query: "After the inspection is complete, is this entire procedure, including the long lines, again necessary to finally get plates?" With a stoicism worthy of a Buddhist nun, the person behind the counter gave a somewhat laconic reply: "Yes." Thus concludeth Day One of the saga of the Department of Motor Vehicles.

It is Day Two of the saga, and Aurora, rosy-fingered dawn, is painting streaks in the eastern sky. Grandson and I are seated in our vehicle, lined up with a number of other temporary licensees at the inspection station behind the Department of Motor Vehicles. The car is to be examined for lights, signals, brakes, and other safety features. It is a procedure I fully endorse, and in déjà vu we once again pass the qualifying exam and find ourselves in the serpentine line clutching a ticket bearing number 712. The signaling panel reads 687, and a quick calculation assures me that the wait today will be only 45 percent as long as on the previous day. Cheered on by that buoyant news, we dash off to a hasty coffee and doughnuts. On return, there are disconcerting numbers on the panel: The line is moving only 45 percent as fast as yesterday, so that the wait is going to be equally long. As we sit waiting, I try to convert these anecdotal results to precise mathematical law. The equation for the Department of Motor Vehicles (DMV) is $D \times M = V$, where D is the density of people in line, M is a motor vehicle department constant, and V is the velocity at which the line moves. Thus, the lower the density of people in line, the slower the line moves.

I cannot begin to imagine the natural mechanism for this seemingly exact mathematical law, but it ensures full employment and guarantees that the staff will be equally busy regardless of how many people show up. This is truly a significant managerial discovery. With wider applications, it would avoid problems of seasonal layoffs and varying staff shortages that plague many industries.

Well, it is now time to review two days, or five hours, with the DMV. I have my shiny new plates and will soon affix them to the car. The mission has been accomplished. But no experience is complete unto itself unless it leads to some future thought, some thrust into the world of tomorrow, and I have a modest proposal, which starts with just a bit of background.

There has been a great struggle in the United States between those who favor strict gun control laws to cut down on everyday violence, mayhem, and murder and those who would like to honor our frontier past with no gun control at all. But surely, everyone can agree that it is as important to register a gun as it is to register a car, and guns deserve the same attention to detail as cars.

Hence, my suggestion: Let us simply turn the registration of guns over to the departments of motor vehicles, applying the automotive requirements. Thus, one would have to appear at the DMV with a title and a bill of sale. This would rid us of stolen and smuggled weapons. The second requirement is that an owner must have liability insurance to receive registration. Thus, instead of overly restrictive government controls, we would let the free-market economy operate and decide who is too great a risk to own a gun. Insurance companies are run by hard-nosed executives and tough, precise actuaries; let them decide who—in regard to character and responsibility—can qualify for liability insurance. Next, force the registrants to go through the same registration procedures as car owners. This would distinguish those who really want a gun from those who are frivolously acquiring one.

In addition, the endless lines at the DMV would separate those stable enough to own a weapon from the hotheads who will blow their cool in the long serpentine lines and glacially slow changing of numbers. They would be carted away for therapy before ever registering a gun. In addition, inspection for the transfer of guns more than ten years old would rid us of unsafe old pieces that occasionally blow up in people's hands.

The idea stands by itself, a monument to simplicity. My grandson thinks I've gotten a little light-headed from all that standing in line.

Delusions of Competence

In 1969, a book entitled *The Peter Principle* arrived on the scene. Written by Laurence F. Peter and Raymond Hull, this seminal work codified the science of "hierarchiology," a discipline that follows from the Peter principle itself: "In a hierarchy every employee tends to rise to his level of incompetence." On reading the small volume in 1970, I was both enlightened and disconcerted. The illumination came in finding an explanation of why so many things go wrong, the malaise from the realization that this is probably the way the world operates. I do not like pessimistic doctrines, even when they are funny and true. I have over the years pondered whether the Peter principle is an inevitable feature of human organizations or represents some aberration of modern-day culture that is subject to attempts at corrective action. After decades of thought, there is some good news to report: I have discovered a mechanism behind the Peter principle. And, as we all know, the elucidation of mechanism is the first step on the way to a cure.

In a hierarchical organization, one performs a task or manages others, or does both. On the bottom rung of the hierarchy, the jobs are all task-related, and there is no managerial responsibility. As one rises in the organization, there is a steady shift to managerial responsibility, so that at the top of the hierarchy the jobs are all management; indeed, they consist largely of managing managers. At the bottom of the hierarchy, the jobs are narrow and specialized, and a highly trained worker can carry out his or her responsibilities with great skill.

As one rises in the hierarchy, one must supervise others and therefore acquire some understanding of other people's jobs and skill in management as well as skill in one's own job. If one is unwilling to learn something about management, one has early in one's career risen to a level of incompetence.

At the next level, one must also supervise workers with different, specialized skills. Thus, one must learn about the tasks these workers perform as well as a more sophisticated set of managerial skills. Incompetence can come at the substantive level of not understanding the tasks to be supervised or not understanding the structure of higher-level management. At each hierarchical level, one is required to supervise workers with an ever-widening set of substantive functions, and one is required to be familiar with a new set of management skills.

Those people who regard each promotion as a learning opportunity and who have the requisite skills may be promoted successfully within the organization. This represents what might be called a normative path of advancement, the way things were viewed before Peter. But the existence of the Peter principle suggests that this often is not what happens.

As an observer of individuals in organizations, I have mused over these problems for many years, and I now think that there is a clue as to why the normative path is not followed. The phenomenon comes under the rubric delusions of competence (DOC). It is really quite simple: When appointed to a job, some people assume that they automatically and suddenly possess the skills and knowledge necessary for the job. Forgotten is the dictum that life is a learning experience, and schooling only prepares one to know how to learn the materials that will arise in later life. The higher the position gained in an organization, the less likely the person is to have the humility to assume the student role that is absolutely necessary for the job. The ego boosting of advancement reinforces DOC.

Let me choose some examples from collegiate life. Suppose, for example, a historian has achieved renown as an expert in the cultural history of Liechtenstein from 1921 to 1927. The individual is asked to become department chairman. Success demands a considerably broadened view of history and an understanding about how to manage unmanageable colleagues. Both of these require either real learning or a residuum of acquired knowledge.

Suppose the same individual were to become a dean. Here the ante is upped, and one is also required to have a modicum of knowledge of subjects other than history, even perhaps a soupçon of mathematics. From a managerial point of view, one must now manage department chairpersons who are even more difficult than history department colleagues, because many of them will be suffering from DOC. And since deans are supposed to know everything, the idea of studying the requisite material seems even stranger and more remote.

Next is the move from dean to provost and an even greater strain on managerial skills, as well as concern with the entire breadth of university disciplines. The range of material is awesome, and the range of personalities is truly mind-boggling. It can be done, and many people have done the job well, but even brief attacks of DOC can be lethal.

Actually, the training in American universities tends to foster DOC in two significant ways. Engineering schools tend to stress technical material, with the idea that managerial skills can be casually acquired. Business schools and MBA programs tend to emphasize the idea that a good manager is a good manager, independent of the nature of the work being carried out at the lower rungs of the hierarchy. Both views are dangerously wrong and contribute to a certain amount of built-in Peter principle. This could be partly remedied within the educational system.

One other feature tends to increase the amount of DOC behavior in an organization. As one goes up the hierarchical ladder, the salary increases out of all proportion to the tasks being carried out. In a capitalist society, people are judged and judge themselves by the size of their paychecks. For the upwardly mobile manager, this has two negative effects: the delusion of competence and an increasing amount of money to spend on doing other things, both of which distract one from learning the substantive and managerial lessons necessary to the new status in life.

Does this analysis provide a cure for the malady? Not really, but it sure does help me understand a lot of behavior I see around me.

Charity for CEOs

A S A WRITER of essays, I struggle to find good news—indeed, I feel something of a moral obligation to write only as much bad news as is absolutely necessary. But the real world is intruding a lot lately, and fidelity to truth is itself a moral value.

So I turn to health care costs and follow a yellow brick road that takes us to the issue of executive compensation at not-for-profit organizations. My thoughts were triggered by an editorial in the *Hartford Courant* (March 19, 1993) that pointed out that the CEO of Yale–New Haven Hospital was paid $518,880 last year. Now, this hospital is a philanthropic institution that regularly goes to the public, hat in hand, seeking funds. Why should Mr. John Q. Public, making, let us say, $50,000 a year, dig deep into his pocket to support the half-million-dollar-a-year lifestyle of an administrator whose duties hardly require unique skills and talents? This is not to denigrate the job of hospital administration, but it hardly requires the towering achievements of an Albert

Einstein, a Leonard Bernstein, or a Michael Jordan. Such hospital executive salaries are clearly part of the high cost of medical care in the United States.

The editorial in the Hartford paper having sounded the alarm, I read with interest an article in the *Washington Post* titled "Probing the Pay at NonProfits" (May 3, 1993). It was noted that the president of Mt. Sinai Medical Center in New York earned a not-too-shabby $799,492 last year. We have just been through the United Way scandal, which started with the exposé of a $400,000-a-year executive salary.

The organizations we have been discussing are not-for-profit; all regard themselves as charities and engage in solicitation. To tie this back to health care costs, the chief executives of major private health care insurers have annual incomes between $1,000,000 and $1,500,000. The obvious conclusion is that CEO salaries are part of the escalation of health care costs. Less obvious is the question of what constitutes reasonable top salaries for executives of organizations that go to the public for funding.

The answer has to be seen in terms of setting national standards for executive salaries in bottom-line (profit) organizations and how they are established. Lower-level recompense is set by upper-level executives. With due diligence, such salaries are kept in line, and a linear rise in salaries accompanies a rise in the organizational hierarchy and responsibility. At the top levels, however, salaries are set by the board of directors, the members of which are both employers of and co-policymakers with the upper executives. In addition, many if not most board members are themselves upper executives in their own right and have a vested interest in the high level of CEO salaries. Thus, positive feedback is established, and upper executive salaries keep escalating out of any relationship to work and responsibility.

For bottom-line organizations, the ultimate limits on this explosive feature are, first, the necessity to keep in the black and, second, the anger of stockholders if dividends fail to meet expectations or fail to materialize at all.

Charitable organizations have the same structure for determining executive salaries, with two exceptions: There are no stockholders, and the response to red ink is to go to the public and ask for more gifts. The boards of directors of these organizations are pleased to go along with these solutions. Hence, charitable organizations develop CEO pay scales that are out of line with the qualifications for the jobs. I would suggest that the entire pay scale is distorted at the CEO level.

Be that as it may, I want to talk about the hospital volunteers. From personal experience, I am familiar with two kinds. One works inside the hospital, helps patients with nonmedical wants such as purchases from the hospital shops, and on occasion wheels patients from one location to another. These tend to be older people, often living on quite modest incomes. A second group of volunteers spends time raising money for the hospital, either by phoning or participating in some other capacity outside of the hospital. They come from a variety of age and income groups but all are interested in their community having a first-rate hospital, and they give of their time to achieve that end. I recall my mother's working as a volunteer in the gift shop of the

Vassar Brothers Hospital in Poughkeepsie, New York. It was her modest way of giving to the community where she had spent her life.

There is clearly a vast difference in the mind-sets of volunteers giving of themselves for the general good and of CEOs whose salaries are a component of the high cost of medical care. The highly paid CEOs in public institutions lend an aura of the "me first" 1980s and make it hard to form a sense of community about the effort. The volunteers are probably unaware of the CEO salaries. Public institutions require leadership standards that are different from those of business, pure and simple.

Will it be possible to find competent management without inflated salaries? Two of the largest charities in the United States are the Salvation Army (net income $1.3 billion) and the Red Cross (net income $1.6 billion). These organizations are led by executives with respective annual incomes of $35,818 and $200,000. These clearly establish a range of values for executive compensation for nonprofits, and one should seriously question any salaries outside this range.

Without a doubt, the executives of our leading organizations should be chosen from among the best and the brightest, but for nonprofit institutions they should also be chosen from among those willing, in some modest way, to give of themselves for community welfare. I don't wish to deliver a sermon, but let us be reminded that avarice does not seem an appropriate qualification in staffing nonprofit organizations.

Laboratory Animals
and Congress

A NUMBER OF YEARS ago, after strong prodding from some members of Congress about the use of mammals in biomedical research, the National Institutes of Health responded through the then-designated Division of Research Resources. That group, which funded many of the nation's animal resources, instituted a study through the neutral offices of the National Research Council. The appointed committee was charged with the investigation of models in biomedical research, with emphasis on nonmammalian organismic models as well as mathematical and computer models.

The twelve-member committee took the charge very seriously. Indeed, one original member withdrew because of an association with an organization that received appreciable support from a high official of a company that was a major supplier of laboratory animals. No taint of conflict of interest was permitted to mar the intended study.

The committee worked over a two-year period and heard investigators from a range of contemporary fields about models in each discipline. A shift of meaning in the concept of models in biologic and medical sciences began to emerge. The committee's initial idea of a model was rather like that of a surrogate—that is, in the sense that a dog with a properly ligated pancreas is a model for a human with diabetes. This has certainly been the conventional idea of a biomedical model, which goes back at least to Claude Bernard and his basic work in physiology. Bernard developed this idea in his highly regarded methodological book, *Experimental Medicine,* first published in 1865.

As the study of models proceeded, it became clear that other features were contributing to the modern view. The universality of intermediary metabolism, the near universality of the genetic code, and characteristics common to all eukaryotic cells were providing such broad intertaxonomic information that useful knowledge did not come from a single species but from a large body of data involving many species.

When one begins the study of a disease nowadays, one first characterizes the state of the organism, then the organ or organs involved, then the specific tissues, then the cells of those tissues, and so on—down to characterization at the cellular, biochemical, and genetic levels. The modeling process then consists of finding analogous or homologous behavior at any of the hierarchical levels involved in the description.

At the organ level, the model system is likely to be in taxa with very similar organs. As one goes toward more cellular description, the range of taxa showing analogous behavior begins progressively to broaden. All of biologic knowledge becomes of potential value in searching for clues to model some specific behavior. Fortunately, the current availability of on-line bibliographic databases allows searching for the best available models at each hierarchical level. Hence, the one-to-one modeling involved in the dog diabetes example is not unique and perhaps not even the principal kind of modeling in contemporary biomedical science.

Without going into all the details, which may be found in *Models for Biomedical Research* (National Academy Press, 1985), some salient points made by the committee can be noted. The committee concluded that the NIH can best contribute to progress across the spectrum of biomedical research through the following actions:

- Encouraging investigators and managers to think of models not necessarily as analogues relating directly to humans on a one-to-one basis but rather as potential sources of information generalizable to the total body of biologic knowledge;
- Encouraging the development of new systems with novel features by supporting research on nonmammalian species, including representatives of taxa that have not previously been well studied;
- Encouraging continued strong support for research on a

small number of intensively studied organisms that are accessible to comprehensive genetic, molecular, behavioral, and developmental analysis;

- Encouraging the development of detailed knowledge about various aspects of the biology of nonmammalian organisms to discover their connections to the rest of the matrix;

- Encouraging the application to and rigorous testing of mathematical modeling in experimental biomedical research;

- Supporting good research without taxonomic or phylogenetic bias, including support for comparative and phylogenetic studies; and

- Continuing to support research using the best mammalian models in cases where adequate nonmammalian models are unavailable.

The Division of Research Resources, now part of the National Center for Research Resources, responded to the report by establishing as one of its components the Biological Models and Materials Research Program. This program now appears as a line item in the NIH budget and gives Congress a chance to respond to its original concerns about the kinds of animal models used in research.

Note that from a formal governmental point of view, everything had been done exactly right in this case. Congress voiced a concern. The agency responded by commissioning a study by an unbiased group. The committee carrying out the study then issued a report, and the agency reacted to the report by setting up a grant program in the area recommended by the committee.

This all sounds very good, but it did not work quite that way. The director's office at the NIH, where budgets originate, did not, and apparently still does not, share the enthusiasm for alternative models—hence the budget for researcher-originated grants in this entire area is only about one one-thousandth of the NIH budget for this type of grant. Congress, with a chance to further its original strong concerns, has not chosen to augment this line item in the NIH budget and hence misses the opportunity to influence research.

Clearly, science has become very big business, and one wonders about the extent to which the legislature should micromanage scientific research by control of line item details. On the other hand, it is difficult to see how a concerned electorate can authorize the expenditure of public funds except through legislative action. In the specific case of significantly expanded models research, the original National Research Council Committee felt that good science was fortuitously in accord with good politics. This is one of the few examples in which the research community and animal welfare advocates are in substantial agreement. That alone should encourage Congress to increase this particular budget item.

[Just to avoid the problem of a hidden agenda, let me note that I chaired the committee that wrote the original models report and that I believe that biomedical research in the United States would be much better off with substantially increased funding of the Biological Models and Materials Research Program of the NIH's National Center for Research Resources.]

History's Black Hole

Some years ago, I found myself at the teachers' residence of a junior high school several miles from the town of Savusavu on the island of Vanua Levu in Fiji. I was conversing with a British volunteer who was teaching English history. I inquired about what I assumed would have been a sensitive subject: "How do you teach about colonialism?" The answer came back: "Oh we don't, we teach the course exactly as we do in England, and we never get to colonialism when we give it there." My surprised rejoinder was, "Never teach about colonialism?" "No," he went on, "by the time we finish the Stuarts and Tudors, there's never any time left for colonialism."

I was being treated to an example of the main problem for historians—the difficulty of sifting through the vast array of material about the past to select those items constituting a coherent, structured account of some civilization or period in time. The process is always difficult, but what are the rules?

I *recalled* this Fijian experience the other day while discussing the history of atomic theory with a group of students. This history can be traced easily in the ancient world from Leucippus to Lucretius, a five-hundred-year period, and then picks up again with the rereading of Lucretius in the Renaissance. In discussing this lacuna with the students, several of whom were history majors, we were once again reminded that intellectual history is taught in the United States as if there were a black hole from the fall of Rome to the Italian Renaissance.

Indeed, little is said about the social and political history of the period from 300 A.D. to 1300. It is rather ignored, giving the impression that the Renaissance arose phoenixlike from the ashes—smoldering for a millennium—of the Classical Age of Greece and Rome. This seems to reflect a northern European perspective and may be part of the gulf that separates English-speaking people from the world of the Middle East and North Africa.

And so we forget that with the fall of Rome, there was conti-nuity of the Byzantine emperors until 1452, and that the Byzan-tine Empire was heir to Hellenistic civilization's rich literature and art. The constant interaction between, and occasional recip-rocating conquest of, the Italian peninsula and Byzantium led to a steady cultural exchange. The Library of Alexandria, that greatest of repositories of scrolls of classical Greece, was proba-bly in continuous use until 646 A.D.

To avoid the millennium in question is to ignore the rise of Islam (622), the expansion of Arabian states (reaching its peak in 714), and the development of major schools of Islamic law, medicine, theology, and philosophy. Most Western scholars are unfamiliar with al-Farabi, Avicenna, Ibn Bājjah (Avempace), and Averroës. These names are not found in our compendia of cultural literacy, although they bridge the gap between the philosophy of classical and modern worlds. Of special interest in the period 700 to 1200 is the flow of ideas among Christian, Jewish, and Islamic scholars.

An example of this breadth of scholarship is the work of the Jewish philosopher Moses Maimonides (1135–1204), who was court physician to Saladin of Egypt. His philosophical masterwork, *Guide for the Perplexed,* was written in Arabic. It was later translated into Hebrew and Latin, and subsequently influenced the major Western religions.

Maimonides' role as a physician also reminds us that Avicenna was known as the prince of physicians. In the great Muslim commonwealth that spread from Persia to Spain, medicine was highly valued, and a number of extant works date from that period. Southern Italy under Islamic influence exemplified the active pursuit of medicine. The medical school at Salerno was established as early as the ninth century. ("Dark Ages," indeed.)

The sophistication of Persian culture during this period is best illustrated in the life of Omar Khayyám of Nishapur (1048–1122). Known to us mainly as a poet, he was a mathematician and astronomer of considerable repute. He built an observatory in Esfahān and was involved in calendar reform. He also was active in the areas of philosophy, jurisprudence, and medicine. His accomplishments speak to the sophistication of the culture in which he worked.

Well, the point has been fairly made. History, as taught in the United States at least, is presented as having a cultural black hole in the Middle Ages. This is a myth that gives a quite distorted view. Why do historians do these things? The answer, I think, is that historical tradition in the United States embodies a tradition of Northern Europe that takes Asian and African history seriously only when it recounts the invasions of crusaders and colonialists.

In short, we have been too busy learning about the Tudors and Stuarts to consider the history of the rest of the world. (Mea culpa, I have not yet mentioned the rich medieval cultures of China, Japan, Korea, India, and southeastern Asia, as well as other places.)

What is the objection to teaching history with large mythic components? To that query there are some evident replies. First, it is not very nice, and we academics should be sensitive to ethical considerations. Second, it devalues the discipline of history and makes it of lesser worth if we have to sort out the real and mythic components of each assertion. And from a practical point of view, the West is dealing with a rapid rise of Islamic culture and influence. To ignore the millennium of 300 to 1300 is to know nothing about the roots of that society's culture. It is difficult to understand societies about whose histories we know so little. In the case of today's Middle East and North Africa, we do so at our own peril.

Is there a solution? I recommend that as a first step, history departments rouse themselves from their dogmatic slumbers and start learning and teaching about the period from 300 to 1300.

Scientific Literacy

ROM MY COLLEAGUES James Trefil and Robert M. Hazen I hear a great deal about scientific literacy. They are the authors of a book on the subject, *Science Matters: Achieving Scientific Literacy* (Doubleday, 1991), and are much concerned about the public's ignorance of the scientific principles necessary to understand modern life. They convinced me that there is cause for concern, and I accepted this message until recent events raised some doubts.

I must begin by pointing out that I rarely watch television—indeed, we have no TV set in our Fairfax, Virginia, apartment. I occasionally watch the Super Bowl and the World Series in neighboring sports bars and sometimes see programs when visiting or traveling. During the past several months I had the opportunity from time to time to see a show I was told was one of America's most frequently watched daytime serials.

The show's format featured a scientific expert seated in a chair having his or her academic credentials, competence, theoretical assumptions, and experimental procedures challenged by a standing inquisitor. This, of course, is the way of science. All results must pass the test of scrutiny by any challenger. Two features distinguished this popular program from actual science. First, the challenges to the expert sometimes had an ad hominem character not usually seen in the meanest of academic confrontations. Second, the questioners, who acted like experts, gave away their lack of familiarity with the material by using syntax subtly different from that used by scientists. At other times the questioners acted like friendly guides, helping the experts tell a story that seemed somewhat contrived. Sometimes two experts offered contradictory stories.

In any case, in one of my random viewings of this strange daytime science show I was treated to a rather detailed presentation of the polymerase chain reaction for amplifying specific DNA sequences. This was followed by a presentation of how DNA sequences operate and produce molecular fragments that can be separated by gel electrophoresis. The level of sophistication seemed to far exceed scientific literacy, and the audience was presented with a large number of complex gels generated from samples of blood found in cars, on socks, on fence posts, and in other strange places.

This reminded me of an earlier episode, the topic of which had been immunology, blood typing, and blood type as a phenotypic marker distributed among various racial and ethnic groups in the Los Angeles area. I can't tell you how surprised I was that such immunochemical detail and the uncertainties of population biology were serving as subject matter for the vast wasteland of daytime television.

Subsequently I saw an episode whose subject was statistical inference based on mixed blood samples from two or more individuals analyzed by DNA amplification and gel electrophoresis. Here the material reached a level of statistical and mathematical sophistication exceeding what I am able to present in my college biochemistry course. Neither my limited knowledge of Bayesian probability nor some dimly remembered discussions of a priori and a posteriori definitions of probability were sufficient to allow me to follow the arguments. I began to wonder how the popular level of scientific literacy had exceeded my own, after my years of study and attempts at understanding. It was downright embarrassing.

There seemed to be no limit to the level of detail that this television program would present. Could it be that Professors Trefil and Hazen have underestimated the scientific literacy of a public that was willing to spend untold hours watching this daytime programming, segments of which were sometimes replayed at night?

This science program reached yet greater heights of sophistication on my last viewing. The subject was the detection of trace amounts of ethylenediaminetetraacetic acid (EDTA), a widely employed chelating agent with both biochemical and commercial applications. Now, this is a compound that I have used and felt I understood quite thoroughly.

But soon the expert and interlocutor were discussing the procedure for derivatization of the compound for gas phase chromatography, the retention time of the derivative, examination of the compound by means of liquid chromatography, and the analysis of the chromatographic components by mass spectroscopy. Next the parent compound, the charge on the ions, and the molecular weights of the derived compounds under varying conditions were considered. At one point the discussion drifted off into whether the peak height or integral of the area under the peak should be used for quantifying the amount of material in the sample. Fine points of sensitivity versus precision were debated with even greater sophistication.

Well, I have no idea of the future of this type of program. This series was referred to as "The Trial" (no doubt in memory of Franz Kafka, who also was interested in long, tedious arguments of uncertain intent). If I were more of a television watcher, I might be better able to put this production in context.

In the meantime, I really must query my colleagues: Can it be true that a populace sadly lacking in scientific literacy could spend endless hours engrossed by such recherché minutiae of science? And what courses could we offer college students not majoring in science that would permit them to understand this material that gave me so much trouble? Perhaps we could try forensic science for nonmajors or biochemistry for poets.

Less Is More

 URING THE twenty years I wrote for the journal *Hospital Practice,* the article that drew the most irate mail was a piece on homeopathy entitled "Much Ado About Nothing." I must say that I was surprised to find individuals who had been through the standard medical training who still espoused a doctrine that I considered just plain silly. Homeopathy is not what I would designate alternative medicine: It is alternative reality.

Thus, I was somewhat surprised when my wife brought home from the pharmacy display area at Long's Drugstore a copy of *The Pocket Guide to Homeopathic Medicine,* put out by Boericke and Tafel, America's oldest homeopathic pharmaceutical firm. I had not expected ever to write about homeopathy again, but the chance to tell the story in the words of the homeopathists themselves was too hard to resist, so I proceed with a condensed version. We begin with a statement of the theory:

Homeopathy teaches us that the symptoms of a disease represent the body's attempt to heal itself. By administering minute amounts of a medicine which produces the same symptoms as the disease, we can reinforce the healing process. Homeopathy works with the body, not against it.

This is a clear and unambiguous statement of the postulate. It contains a rationale for dealing solely with symptoms rather than focusing on underlying causes or etiologies. We next deal with how homeopathic medicines are made:

The first step is to prepare a mother tincture (similar to an herbal tincture) from the raw material. The mother tincture has a potency of 1X.

Next comes a process exclusive to homeopathy called potentizing, which consists of diluting and succussing (shaking vigorously in a prescribed manner) the medicine.

To raise the potency from 1X to 2X, one part of the 1X solution is mixed with nine parts of alcohol and succussed. To raise 2X to 3X, one part of the 2X solution is diluted with nine parts of alcohol and succussed. The process is repeated until the desired potency is achieved.

Note the implications of the preceding: the higher the potency, the less of the original material in the preparation. This is contrary to usual usage but well defined.

The material is dispensed as tablets or pellets. The pocket guide does not indicate how one goes from an alcohol solution to a tablet or pellet and how much of the liquid is used per unit of solid dosage and what else is in the solid form. This is important missing information for any evaluation of the procedure.

Also contained in the guide from Longs is a Potency Chart, which contains three categories: low (6X, 12X), medium (30X), and high (200X). A "potency" of 24X will contain less than one molecule per quart of solution, a quantity far below the ability of chemists to measure and at least ten orders of magnitude below the level at which any noninfectious particle has ever been measured as having any physiologic effect.

A "potency" of 30X means an alcohol solution with one molecule of the active material per one thousand tons. Your chance of getting one molecule of the active substance in a tablet or pellet is in the order of one in a hundred million.

A "potency" of 45X corresponds to one molecule in all the oceans of the world, and a "potency" of 200X corresponds to a concentration something like one molecule in the observed universe. Yet a look at the potency chart will tell you that this "potency" should be used by practitioners only.

Wishing more detailed information, I went down to a local pharmacy that dispenses homeopathic "remedies." They carried a different brand than Long's, one supplied by Standard Homeopathic Company of Chicago. Reading labels, I noticed many of the ingredients were at potencies of 1X to 6X. Now, 1X, 2X, and 3X did not seem sufficiently dilute to be homeopathic, but rather a case of natural product therapy. However, that's what was printed on the label.

I noticed that one of the nostrums was labeled "No. 23 Insomnia. *Active Ingredients:* Hydroscyamus 3X HPUS (0.00016% alkaloids). Ignatia Amara 3X HPUS (St. Ignatius Bean 0.0012% Alkaloids), Kali Phosporicum 3X HPUS (Potassium Phosphate). *Directions:* Adults: 4 tablets at bedtime, then hourly if wakeful." A friend had been telling me of problems falling asleep, so— always the experimentalist—I shelled out $5.29 and purchased 250 tablets.

Given the theory of homeopathy that less is more, I was surprised at the four-tablet dose, so when my friend reported no effect, I recommended cutting the dose to two tablets. When this failed I suggested one tablet, and finally breaking a tablet in half. Alas, I was of no help to my friend.

But on thinking about the theory of homeopathy, I decided to experiment on a cure for insomnia. I made a cup of very strong espresso coffee and diluted it ten to one, being careful to succuss it. A cup of the dilution didn't work, so I tried espresso at 2X (one to a hundred), and that also failed. I am now out to 10X and can report no success. If it turns out to work at, let us say, 12X, I can convert a cup of espresso to approximately thirty billion gallons of remedy. I'll be able to put the whole world to sleep, and at $5.29 a shot, I'll be able to buy out Ross Perot and George Soros in cash.

Why am I making light over a point of view that so many take so seriously? In part it's because I don't know how else to respond to an allegedly scientific theory that at its core is in methodological violation of the epistemological bases of that science.

Rationality has its bounds, and it is not true that anything goes. There is the innovative, the imaginative, the idiosyncratic, the ill-construed, and then there is the inane.

What I do take seriously is the Food and Drug Administration and what it is doing to guarantee the efficacy and safety of homeopathic pharmaceuticals. The writings I have quoted constitute labeling, and I want to know how the efficacy has been established to the satisfaction of the FDA. Are they going to allow me to sell espresso diluted a trillion times and maintain relief for insomnia? If not, what are they doing about the homeopathic remedies sold around the country?

VI

Commentary

Wheels

I ARRIVE IN Albuquerque and confidently march up to the car rental agency to claim the compact I reserved. The clerk, operating with a computer system that is down, shuffles through the requisite papers and asks, "Will you take a Lincoln Town Car?" I must look crestfallen, because she hurries to note, "It will be at the same price." Not wanting to make a fuss. I find myself shortly thereafter sitting in a substantial blue-gray vehicle.

Before continuing with this story, I must discuss my philosophy of transportation. I normally drive a 1986 Ford Escort station wagon. It carries its share of dents and scrapes, as well as a red plastic backup-light shield with jerry-built repairs. It is not a car that puts me in great danger of violating the speed laws. Indeed, when carrying a passenger on hot days, I have been known to turn off the air conditioning to make it up a steep hill. It is economical, reasonably reliable, environmentally acceptable transportation. And within my current Thoreauvian, suburban philosophical frame of mind, it is what I require as a mode of transportation. Thus, my large, vast, comfortable rental vehicle does not fit easily into my current lifestyle.

It is not as if I have never driven a large car before. Many years ago I had left my Rambler parked at the New Haven railroad station and had gone into New York City to do some consulting. The sales manager of the company I was working for invited me to return to New Haven that evening in a rented copper-tone Rolls-Royce that was being used in a promotional activity.

On arrival in New Haven in the early evening, my companion and driver said, "Why don't I get off at the hotel? You take the car home, pick me up in the morning, and I'll take you to the station to get your car." So I drove along to the first pay phone, called my wife, and said, "Let's go visit a few friends this evening."

At breakfast the next morning I suggested that my oldest son, who was then in junior high and very interested in cars, might want to look in the garage. He returned a few minutes later with the smart-kid comment, "Gee, that was a shrewd trade you made for the Rambler!" After that, there was no backing out—I began my workday by driving the children to their various schools in the prestige machine.

I then drove to the downtown hotel, picked up the sales manager, drove to the railroad parking lot, and was finally back into my comfortable Rambler. You will note that I drove a Rambler until I graduated to status cars like used Ford Escorts and Chevrolet Cavaliers. I present this background so you will appreciate that I am not totally inexperienced in this business of driving large, expensive luxury machines.

A few minutes from the airport, and I am navigating my new-found extravagance north on I-25 to Santa Fe. The ride is as comfortable as the advertisements claim, and I am overcome by a strange sense of well-being. At first, I attribute it to the new car. But no, something else is operative. It is one of those days when the New Mexican air is so clear and crisp it seems nonexistent. The unfiltered light makes objects seem luminescent. I remember reading that Georgia O'Keeffe moved to New Mexico in part because the light was so favorable for painting. This morning I have a precise sense of what she meant, even though I have never painted. If the car came with an easel, canvas, and set of oils, I think I would pull off to the side of the road and begin a new career.

In such comfortable and ethereal surroundings, it takes an act of will to keep the speed down, and I struggle to make the effort. I enter Santa Fe and park in the lot of the Santa Fe Institute. As I get out of the car, I see a colleague. An explanation seems in order. For a Thoreauvian suburban driver of a 1986 Ford Escort station wagon, a large, shiny Lincoln Town Car comes with a certain measure of guilt, so I elaborate on how I am driving this vehicle as an act of Budget Rent-a-Car, and not of my own free will. Such feeble attempts to remove guilt are never very credible.

For the next few days, my freedom of motion is somewhat inhibited. The shiny car is so long and so wide, and parking spaces are so small. I move comfortably from the motel parking lot to the institute parking lot, but other trips around town are less relaxed. I begin to wonder if I am suitably well dressed for the car, or whether I should put on a necktie for driving. (Fortunately, Santa Fe is a quite informal town.) In addition, Budget has neglected to include the owner's manual and many operating buttons on the seat handle, steering wheel, and dashboard are something of a mystery to me.

A worry begins to emerge in my psyche. I'm beginning to feel a certain disdain for people in lesser vehicles who do not give me the right of way and offer other appropriate acts of respect.

In any case, it is time to return, and the trip to the airport is on a somewhat overcast afternoon, so there is no distraction from the artist's light, and I am totally into the existential experience of driving and thinking about motor vehicles and societies. All sorts of strange thoughts pass through my head. When I arrive at the airport, my mind is set on the appropriate response to the events of the past few days.

"Please don't give me one of these again," I request, returning the keys to the clerk. "I'm returning it just in the nick of time. All the way back, I kept complaining about the capital gains tax situation and regretting not having voted Republican. Another few days in this vehicle and my life will be irrevocably altered. I kept wondering why I'm not flying first class. When I reserve a compact car, please try to give me a compact. Actually, a subcompact will do."

Civic Duty

I T WAS EARLY one Monday morning on the fourth floor of the Federal Building. In the small foyer of the Internal Revenue Service offices, a few of us sat anticipating the call. My fellow citizens appeared to be somewhat disconcerted, if not downright jittery. After a short time I was called in to meet my auditor, a cheerful looking, mild-mannered woman.

As she began to shuffle through some papers, I thought a little conversation was in order: "You know, I'm pleased to be here for this routine audit," I said. "I regard it like jury duty—one of those obligations of citizens that are not especially pleasant in themselves but necessary. Without the work you auditors do, I'm sure our national system of raising money by taxation would collapse."

She looked at me quizzically and then broke into a smile, saying, "Well, you've made my Monday morning!" I was happy that my link to the government of the United States of America was also happy, and we proceeded to some dull matters of numeracy. Two hours later, we were still in perfect agreement, this time about the books. Two compensating errors had been found. In one I had not understood the instruction properly, and in the other I had (you will please keep this from my students) made a numerical error. As the errors did not completely balance out, we agreed that the Internal Revenue Service owed me $300. I did not press for immediate repayment. My auditor had been so pleasant, I thought it would have been gauche to talk of cash on the barrelhead. After all, if we can't trust the IRS, whom can we trust?

As I drove home, I reflected on my simile about audits and jury duty and reminded myself that I had never actually served on a jury. A number of years earlier, I had received a notice to appear for jury duty at a Connecticut state court, but there had been a problem. Although I was available for the preliminary jury selection meeting, on the date the trial would begin I was scheduled to be in Houston for a meeting of the NASA Planetary Biology Subcommittee. The subcommittee was to make the final decision on which experiments would accompany the *Viking* lander exploration of Mars. Thinking it would be a routine matter, I called the court clerk; I reasoned that such a cosmic purpose would surely justify a postponement until the next jury selection.

I was wrong. The clerk informed me that only the judge could issue excuses and that it was necessary to appear in person on the day the group of potential jurors was to convene.

Thus, I appeared at court in Waterbury on the appointed day, with about three hundred other panelists. When the judge appeared, he directed all those who wished to be excused to go upstairs to room B250. About one hundred of us did as directed and sat in the smaller courtroom. About forty-five minutes later, the judge appeared and began calling us to the bench one at a time, summoning us by our towns of residence in alphabetical order. As a resident of Woodbridge, I knew it was going to be a long day. He started with Ansonia, and there are 190 cities and towns in Connecticut.

The excuses were predictable and the judge was tough. The most frequent excuse had to do with the care of a sick relative, usually a parent. The judge asked probing clinical questions to see if the care was indeed required. Other excuses involved financial exigencies, transportation problems, and illness on the part of the potential juror that would render service on the jury a threat to his or her health. His Honor did not excuse everyone.

I was the next-to-last venireman to be questioned. I had lost a day of work. The judge was tired and a bit grumpy. As I began to tell my story, he perked up. Here was an excuse he had not heard, and he wanted to make the most of it. He asked me for details about the Planetary Biology Subcommittee and why they needed me. A difficult question! If I were too modest, then it would appear that I was not needed and I would miss a meeting that was the culmination of three years of work. If I were too arrogant in describing my role, he might react negatively. I had to walk a narrow path, and so I did.

Finally, he said, "All right, I'll let you go this time, but I never want to hear that excuse in my court again." I thanked him and left. I was never called to jury duty again, but I have paid my civic dues by a routine IRS audit, so I can relax. I don't, however, really know if my presence in Houston made any difference in the experiments that were selected. We had reached a consensus.

Viking lander was launched and did gather interesting information about the Martian surface, including little indication of life. The mission was a triumph of human exploration, and I am pleased to have been associated with it. Future lander missions will doubtless take place. If I am involved in these endeavors, I can only hope that they don't overlap with another call to jury duty in Waterbury—or a tax audit, for that matter.

Mental Calisthenics

SOME YEARS AGO I received a questionnaire from a foundation looking into the question of the best undergraduate training for predivinity students. The accompanying letter indicated that this was being addressed to one thousand faculty members chosen at random. Most of the time I ignore such studies because of a feeling for the difficulties in formulating the questions so as to get reliable and useful information. On this occasion, the question was sufficiently different and engaging that some follow-up thought was appropriate.

My recommendation was physics. The theoretical aspects would give students training in dealing with the abstractions necessary to theology, while the experimental aspects would provide an anchoring of thought in the everyday world where results are indeed testable. These conclusions were duly sent off, and I never received a reply with the results of the study. Perhaps I had confused or offended the questioners.

In any case, the issue of undergraduate and high school core curricula is ever with us, asking for the best choice of subjects for preparation for this or that career. So on occasion, as another generation matures, there comes a time to reexamine the question of choice of subjects. A few universals have emerged in my own thinking that can perhaps provide some basis for arguing these points.

I am firmly convinced that every high school student should study physics. The reasons include those I gave for the aforementioned study, but they are somewhat broader. Physics provides the basis for understanding chemistry, biology, geology, and the engineering disciplines. As such, it allows one to deal with the phenomena of everyday life, not as intellectually isolated but as part of a network of understanding that has emerged over the past three hundred years. A large subset of phenomena are effectively demystified by an understanding of elementary physics. As a result, a number of repairs and other manipulations of everyday objects, as well as responses to common situations, become easier and more intuitive.

From a more philosophical point of view, physics, while limited in its domain of applicability, provides a sense of the rationality of phenomena within its scope that is not available in any other endeavor. This, of course, does not mean that other disciplines need to imitate physics. Attempts to do this in the social sciences have often been intellectually disastrous; nevertheless, it is well to have a paradigm subject that can, in certain areas, be thoroughly understood.

The study of high school physics is going to require a certain background in mathematics. This prerequisite is, of course, part of the recommendation of a year of physics. Mathematics has other values all its own.

For most students, I would add Latin to the list of core courses. The reason for Latin is somewhat less universal than that for physics, but nonetheless I find it persuasive. First, the Romance and Western Germanic languages, including English, are the central carriers of Western culture, which is the major framework of current American civilization. Latin, at the literary core of the Romance languages, is the historical root of Italian, Spanish, Portuguese, French, and Romanian, as well as a number of minor languages. For students who start with a knowledge of English, a knowledge of Latin positions them in Western culture.

The religious belief system that shaped Western Europe and the Americas was articulated entirely in Latin from Saint Augustine to the Reformation. Even the physical and biological sciences were written in Latin, from the *De rerum natura* of Lucretius to sometime after the *Principia* of Newton. The roots of modern education, the trivium and the quadrivium, are known to us by their Latin names. The Renaissance was in part the latinization of secular learning in Europe.

Of course, it might be argued that other languages have had equal influence on global culture, and I would agree. Arabic and Chinese could certainly be viewed as equal in breadth. But the students I am thinking about are in the contemporary United States, where the dominant cultural elements, including science, are as described above.

Because it is primarily a written language, with structures that have been fixed for some time, Latin serves as an introduction to thinking about language—linguistics. In the modern, information-driven world, linguistics becomes more and more important.

One other feature is held in common by physics and Latin: Both are highly disciplined subjects. Much of modern high school education seems to lack structure and responds to a wide-ranging social agenda. It is important to have some portion of the curriculum that can be learned in a systematic manner and that admits of understanding by learning the underlying structure. Physics and Latin at the high school level can be learned by dedication and commitment. Complex social issues, however, carry a certain amount of ambiguity, and literature is highly individual. While physics and Latin are social activities to the core, they are relatively free of such uncertainties. They both depend upon a consensus of participating individuals who have established an understanding.

In athletics, music, and a number of other areas of interest, our students understand practice, discipline, and repetition. Mental calisthenics are also important. They are provided by Latin and physics.

Wizardry

ALAS, I must confess that, having seen the movie version of *The Wonderful Wizard of Oz* four or five times while my children were growing up, I had never stopped to think of the deeper meaning of this story by L. Frank Baum. I wonder how many other opportunities to probe the world in depth have, similarly, passed by me unnoticed? Indeed, I suppose I would have remained in ignorance of this particular instance except for a casual comment by an economist friend, pointing out that this delightful children's story was a political allegory on the monetary system.

That statement is the kind of challenge that seizes me and will not let go. I suppose that my reference librarian is right when she refers to me as "a bibliographic database junkie." It's not easy to go public with one's faults, but best to confess and go on with one's life.

The challenging statement about Oz first came to me in

Lahaina, Hawaii, and I have been pursuing the information in a nationwide trek, beginning at the Lahaina Public Library, continuing at the Santa Fe Public Library, and going on to the George Mason University library. I am now at the Cowles Institute of Economics at Yale, where a kindly librarian has admitted me to practice economics without a license. Before me is the *Journal of Political Economics,* volume 98, opened to page 739 and the object of my long quest: the article *"The Wizard of Oz* as a Monetary Allegory," by Hugh Rockoff. Surely, I had come to the canonical work on this subject.

efore launching into a review of Rockoff's conclusions, let me briefly report on my findings while on the road. First, I found that Oz had entered the mainstream of twentieth-century American critical literary life with writings by Martin Gardner and Russell Nye, and had indeed been considered from points of view that ranged from those of populism to those of psychoanalysis. The movie, a classic in its own right, has also been extensively scrutinized in terms of cinematography and dramaturgy.

At least four theories have been proposed about the origin of the word *Oz,* the name of the Emerald City. The explanation most often quoted is that Baum, seeking a name for the city, looked up at his file cabinet and saw a drawer labeled O–Z. A second theory is based on the notion that Charles Dickens, who was Baum's favorite author, had the nickname Boz (in his *Sketches by "Boz,"* 1836). The reader will forgive my forgoing a critical study of Dickens to track down this proposal. A third view emerges from Baum's familiar comment about liking a story with a lot of *oh*s and *ah*s. The most scholarly hypothesis is that Oz is named after Uz, the land of Job in the Old Testament. Alas, this proved to be a scholarly conundrum, for Uz as a geographical place name occurs three times in the Old Testament, designating three different places (see Job, Jeremiah, and Lamentations). There are clearly no limits to how far one can go with this kind of scholarship.

Before moving on to the monetary allegory—in which Oz stands for ounce of gold—a little background is in order. In 1873, the United States stopped coining silver dollars, making silver secondary to gold, which was controlled by East Coast bankers who supported the gold standard. Populists objected, and the arguments lasted for many years, highlighted by William Jennings Bryan's "Cross of Gold" speech and terminated by the Republicans' passing a Gold Standard Act in 1900. *The Wonderful Wizard of Oz* was written in 1899.

There is no way to be sure that Baum intended an allegory, for no record exists on that point. Rockoff acknowledges this limitation and proceeds to build on the work of others to draw the analogies. I must confess I have a few tidbits of my own to add to the presentation in Rockoff's fine paper.

Dorothy from Kansas is Middle America, and Oz, the Emerald City where everyone wears green glasses, is taken to be Washington, D.C. The Wicked Witch of the East, representing the Eastern financial establishment, dries up completely, leaving silver shoes or the silver component of the bimetallic standard. (Note that later, when Dorothy destroys the Wicked Witch of the West by dissolving her with water, all that is left is a gold hat. Could that be panning for gold?) The Scarecrow and the Tin

Woodsman are taken to represent the farmer and the laborer, and the Cowardly Lion was William Jennings Bryan. This is the populist foursome: the farmer, the laborer, the Middle American, and Bryan.

The treacherous yellow brick road is the gold standard, and in this version the name Oz is taken, as we have noted, from the abbreviation for ounce of gold. Who, then, is the Wizard? He appears not to be William McKinley, the president, but Marc Hanna, the chairman of the Republican National Committee, "who speaks through various figureheads and adheres to a purely Republican world view."

The monetary policy issue is again revived when the Wicked Witch of the West is dead and the Tin Woodsman is given an ax with a golden handle and silver blade and a silver oilcan inlaid with gold, Toto and the Cowardly Lion are given gold collars, and the Scarecrow is given a walking stick with a head of gold.

Well, that is the story of Oz according to an economics model. Is it true? We shall probably never know, as is the case with most economics models. However, I am sanguine about the whole situation. If I lack certainty about the allegorical interpretation (and I do), I have learned a bit about U. S. economic history. Best of all, I have been led to read the original *The Wonderful Wizard of Oz* by Baum. That is reward enough in itself.

Oh, I almost forgot. Toto, the name of Dorothy's dog, is supposedly taken from teetotaler, a minor party that may have held strange views about drinking but was right with the populists on important ideas such as silver.

Digitoma

ONE OF THE most subtle features encountered when describing a new construct is choosing an adequately descriptive title. When the construct involves a social pathology, one tends to turn to the rich vocabulary and complex syntax of medicine. Thus I must hasten to state that digitoma, the subject of this essay, is not a tumorous state of the toes, nor a profuse sprouting of fingers—rather it refers to the explosive growth of indexing numbers or, more accurately, the number of markers in an index number.

I first began to think about the unrestrained growth in the length of indices when the post office opted to add four digits to the five-digit zip codes. In a base-ten system, the nine symbols will describe a billion areas with their own zip code. Since the area of the United States is about 3,600,000 square miles, and much of it is unoccupied, it seems strange to have a zip code for every 0.0036 square miles, or about 100,000 square feet. Having a separate zip code for every plot 317 feet by 317 feet, whether occupied or not, seems excessive. Viewed differently, there are four zip codes for every man, woman, and child in the United States.

While this thought intrigued me, I could not place it in the cosmic context that it seemed to deserve. Is the post office anticipating a population explosion that will require a billion zip codes? The very idea is staggering. What unanticipated contingencies are driving this expansion of numbers? I placed the idea on the back burner, where it has sat these many years.

Then an event of not very great profundity brought the matter back to mind. We had stopped for a red light, and the car behind us didn't stop quickly enough. The damage to our vehicle involved the rear bumper and its assembly. There was no question as to who was responsible. I phoned the covering insurer, arranged an appointment with their adjuster, and was given a case number. All went smoothly enough, but one aspect of the number puzzled me; it was sixteen digits long. Since the mathematics was to the base ten, my case was being distinguished from 10^{16}, or 10,000,000,000,000,000 other claims. One wonders what future volume of claims the insurer was anticipating.

While I was wondering what to make of these sixteen digits, life went on, and I found myself in a distant city phoning home. The entire dialing operation took thirty-six digits: one to get an outside line, eleven to get my phone company, eleven to get to my destination, and thirteen to charge my call. The thirty-six digits in this transaction represent 10^{36} possibilities. Now 10^{36} is a big number. It would allow me, for instance, to index every molecule in the body of every man, woman, and child in the United States.

I can remember the days when my father's business had a three-digit phone number and our home had a four-digit number. My wife tells me that her grandfather's phone had a two-digit number. Those were, of course, the days of telephone operators and mechanoelectrical coupling. The size of the board and the number of plug holes dictated limitations. The invention of business machines, computers, and electronic switches greatly expanded the number of digits that can be used in a transaction. Given the ability to handle an almost arbitrary number of digits, we are tempted to carry all sorts of marginal information in our transactions. Being blessed by these wondrous ways of handling information, why do I express negative feelings? The answer is that in the loop with all the silicon chips are human beings who must from time to time deal with all these numbers.

To see the pathological features of digitoma, we must remember that to err is human and to forgive divine. However, computers are very far from divine and are extremely unforgiving of wrongful entries. Thus, to make our humanity more metric, let us assign p as the probability of human error when using a single digit in sequence. The probability of not making an error is $1-p$. If I use a sequence of n digits, the probability of getting the whole sequence correct is $(1-p)^n$. For very small error rates, that is for low values of p, the previous expression can be approximated by $1-pn$, and the probability of making a mistake is pn. As n gets very large, the burden shifts to humans to make p very, very small, but to err is human.

There is a branch of information theory that has to do with optimally encoding information, and I am sure that, if it were applied to the problem of digitoma, we could appreciably reduce the number of index and code markers for various transactions. That no significant effort has been made suggests that we would rather modify humans to fit the machines than the other way around. I imagine that this involves inordinate time spent by humans punching numbers into computers and frequently correcting those numbers. I am acutely aware of this when standing in line at an airline counter watching simple transactions which involve a total of over one hundred markers.

Well, it is clear that we live in a digital age, and I had better get my act together and stop dialing the wrong number. If I can get my error rate down to 0.01, then I can dial ten-digit long-distance numbers correctly 90 percent of the time. Gracious, what would Thoreau have said?

The Closer

BASEBALL sentimentality is a bit of Americana that has my generation firmly in its grasp. To be sure, after the 1996 strike, it was something of a love-hate relationship. It is painful to realize that those icons on the baseball cards are afflicted with greed along with the rest of us. I have made my peace with these conflicting emotions by attending minor league games at such diverse locales as the home parks of the New Haven Ravens, the Prince William Cannons, and the San Jose Giants.

For those sociologists who wish to study middle America, I recommend for observation the fans at minor league games. When they stand at the seventh inning stretch and sing "Take Me Out to the Ball Game," all the existential angst of today's drug- and crime-ridden society temporarily vanishes in a haze of nostalgia. I remember that my mother had the sheet music to that song; a salty drop courses down my cheek.

However, even at minor league ball games I have come to see

a dark cloud on the horizon, in that management practices rob the game of a certain natural ebb and flow. Picture a lively game at the beginning of the seventh inning, with the home team leading 2 to 1. The home pitcher has worked six innings and done a good job. Nevertheless, the manager calls in a relief pitcher, who promptly gives up three runs, and defeat is snatched from the jaws of victory. The manager has violated the first law of sanity: "If it ain't broke, don't fix it."

My memory of baseball goes back a long time, even to the days of Lou Gehrig, Carl Hubbell, and Mel Ott. The doctrine of the fragile pitcher is a recent one. I can even remember an end-of-the-season doubleheader when Don Newcombe of the Brooklyn Dodgers started both games in a heroic effort to win the pennant. The attempt failed, but what nobility!

The classic reason for removing a pitcher and substituting a reliever was quite simple: The opposing team was being too successful at scoring. No one thought about how many pitches the hurler had thrown. No one even counted, so that the metric of total number of pitches was unknown. The modern doctrine is that at some mystic number between 100 and 120 pitches, the human body fails and falls into some inferior baseball mode. Where does that number come from? Was it calculated on the basis of careful measurements taken by exercise physiologists? I doubt it and instead suggest that among a group of very superstitious managers the number was obtained by some sort of contemporary cabala (number magic of the Middle Ages). In any case, the idea is widely held that the magic number of pitches, x, is a universal constant, independent of the age, physiologic state, or free agency status of the pitcher. This clearly seems like an extreme hypothesis, but so it goes.

While the "magic number" doctrine was being institutionalized, the "closer" rule was being developed. This rule states that, no matter what the state of the game, a pitcher called "the closer" must appear for the ninth inning. I believe I have some insights into this practice. Some years back the Oakland Athletics acquired a remarkable pitcher named Dennis Eckersley. It was found, experimentally, that he had the astonishing skill of being able to emerge from the bullpen and get three outs very expeditiously before any runs were scored. And Eckersley could accomplish this feat every day. So whenever the Athletics were not hopelessly behind, he was called upon to pitch the ninth inning.

Other managers witnessing this phenomenon wanted closers of their own. The practice spread to the minor leagues, and each team designated a pitcher for this task. The problem is that there is only one Dennis Eckersley, and other pitchers cannot match his marching from the bullpen and delivering three outs. Alas, even Eckersley can no longer perform this feat with certainty. The more likely scenario is that a team is leading 3 to 2 going into the ninth, and the manager calls for the closer. This false savior, un-Eckersley-like, yields a single, a walk, and a home run, and defeat is once again snatched from the jaws of victory by the score of 3 to 5.

Why then do managers, who are reputed paragons of independence, embrace these pitching paradigms? First, it's an American thing, or a human thing—not a baseball manager thing. All of us apparently follow unquestioned rules of our particular discipline. These may or may not be good rules, but the acceptance seems to be part of the game, and few, even among the most independent-seeming people, want to buck the tide of perceived wisdom.

Second, the manager's ego is on the line. If he sits back on a bench, chaws tobacco, gets fat, and lets the pitcher win the game, people will begin to ask, "Why do we need a $275,000-a-year manager to watch the team win?" So there are things a manager must do. He must periodically argue with the umpire and get thrown out of the game three times a season. He must occasionally growl at players and make a big deal about right-handed pitchers and left-handed hitters, and vice versa. There are a number of rituals that are incumbent on a manager, and these are universally observed.

Of course, every field has these procedures: the obstetrician who interferes in a normal delivery, the surgeon who doth excise too much, the lawyer who sues too often, and the scientist who publishes the same paper in three different places. The necessity for self-justification, the importance of doing something when nothing is called for, the inability to refrain from fixing that which is not broken are characteristics of professionals in many fields. In baseball these are more apparent because the job is performed in public, and everyone can second-guess the manager.

However, in spite of these quibbles with the managerial profession, I shall continue to watch minor league baseball. The joy of rising in the seventh inning to sing is just too great to endanger by disciplining the managers. Besides, managerial protocols don't influence the outcome, because in any given game both opponents make the same moves, so that there is offsetting silliness. But someday a manager will appear who will break the paradigm, and real excitement will follow. I hope I'm there to see it.